Statisti
for the Ter

FOURTH EDITION

Statistics for the Terrified

John H. Kranzler
University of Florida

PEARSON
Prentice
Hall

Pearson Education International

Executive Editor: Jeff Marshall
Editorial Director: Leah Jewell
Editorial Assistant: Jennifer Puma
Senior Marketing Manager: Jeanette Moyer
Assistant Managing Editor: Maureen Richardson
Production Liaison: Fran Russello
Manufacturing Buyer: Sherry Lewis
Art Director: Jayne Conte
Cover Photo: Photo Link/Photodisc Red/Getty Images, Inc
Director, Image Resource Center: Melinda Patelli

Manager, Rights and Permissions: Zina Arabia
Manager, Visual Research: Beth Brenzel
Manager, Cover Visual Research & Permissions: Karen Sanatar
Image Permission Coordinator: Annette Linder
Photo Researcher: Kathy Ringrose
Composition/Full-Service Project Management: GGS Book Services/Chitra Ganesan
Printer/Binder: RR Donnelley & Sons Company

Credits and acknowledgments borrowed from other sources and reproduced, with permission, in this textbook appear on appropriate page within text.

Pearson Education LTD. London
Pearson Education Singapore, Pte. Ltd
Pearson Education, Canada, Ltd
Pearson Education–Japan
Pearson Education Australia PTY, Limited

Pearson Education North Asia Ltd
Pearson Educación de Mexico, S.A. de C.V.
Pearson Education Malaysia, Pte. Ltd
Pearson Education, Upper Saddle River, New Jersey

10 9 8 7 6 5 4 3 2 1
ISBN : 0-13-232886-0

To my wife, Theresa,
on account of whom I smile a lot

CONTENTS

APPENDICES

PREFACE

Fear of heights, fear of public speaking, fear of snakes—virtually everyone is afraid of something. *Statistics for the Terrified* (4th ed.) is a user-friendly introduction to elementary statistics, intended primarily for the reluctant, math-anxious/avoidant person. Written in a personal and informal style, the aim of this book is to help readers make the leap from apprehension to comprehension of elementary statistics. *Statistics for the Terrified* is intended as a supplemental text for courses in statistics and research methods, as a refresher for students who have already taken a statistics course, or as a primer for new students of elementary statistics. Millions of people have math anxiety—yet this is rarely taken into consideration in textbooks on statistics. This book presents self-help strategies (based on the cognitive behavioral techniques of rational-emotive therapy) that help people manage their math anxiety so they can relax and build confidence while learning statistics. *Statistics for the Terrified* makes statistics accessible to people by first helping them manage their emotions and by presenting them with other essential material for learning statistics before jumping into statistics. After covering the essentials required for the journey into statistics, the remainder of the book presents an introduction to elementary statistics with a great deal of encouragement, support, step-by-step assistance, and numerous concrete examples, without lengthy theoretical discussions.

ORGANIZATION

This book is divided into four sections. Section I—"Essentials for Statistics"—consists of three chapters. The first chapter introduces the text and presents strategies for studying statistics; the second discusses self-help strategies for overcoming math anxiety; and the third presents an overview of SPSS, one of the most powerful and widely used statistics software programs. Section II—"Describing Univariate Data"—contains chapters on frequency distributions, descriptive statistics, the normal curve, and percentiles and standard scores. Section III—"Correlation and Regression"—consists of chapters on correlation coefficients and linear regression. Section IV—"Inferential Statistics"—contains four chapters on understanding inferential statistics, the t Test, analysis of variance (ANOVA), and chi-square. The final chapter summarizes the text and congratulates the reader on a job well done.

CHANGES FROM THE THIRD EDITION

The format for the previous three editions of *Statistics for the Terrified* was to introduce statistics concepts and methods in an informal and straightforward manner, then walk readers through each step of the process required for conducting statistics or creating charts and graphs. This often involved computing statistics with handheld calculators and looking up critical values in tables to determine statistical significance. The widespread use of statistical software such as SPSS has changed the way that statistics is taught and learned, however. *Statistics for the Terrified* (4th ed.) was written with this in mind. While the format and topics covered in the text remain largely unchanged, in this edition readers are shown how to enter data, compute statistics, and interpret output, using SPSS. Step-by-step instructions with multiple screen shots are provided throughout to assist the reader in learning each concept. In addition to this major change, several other smaller changes were made in the fourth edition. First, the basic math review in Chapter 3 of Section I is presented in the appendix. In its place is a chapter with an overview of how to use SPSS. Tables of critical values and the list of formulas also were removed from the appendix. Finally, to further lighten the content and make reading this book more enjoyable, humorous cartoons and jokes about math and statistics topics were added throughout.

ACKNOWLEDGEMENTS

My father, Gerald D. Kranzler, was a professor of counseling psychology at the University of Oregon. His primary scholarly interest was rational-emotive therapy, but he also taught an introductory statistics course. He is the principal author of this book. My father passed away in 1994, shortly after publication of the first edition of *Statistics for the Terrified*. Dr. Janet Moursund was his coauthor for the first two editions of the book. I have revised a great deal of their book in the last two revisions, but I tried to retain their style and approach, which are the heart and soul of the text. I would like to thank Eric Rossen, and Krista Schwenk doctoral students in School Psychology, for their careful review of the text. Any errors or omissions are mine, of course.

I also wish to thank the following reviewers: Ron Salazar-San Juan College; Melanie Narkawicz-Tusculum College; Olivia Yu-University of Texas-San Antonio; James Knapp-Southeastern Oklahoma State University; Lisa Waldner-University of St. Thomas; and Stacy Ulbig-Southwest Missouri State University.

I also need to acknowledge the following websites, from which I borrowed many of the math and statistics cartoons, jokes, and quotes contained in this edition:

home.earthlink.net   www.sciencecartoonsplus.com
www.csun.edu   www.borg.com
http://www.xs4all.nl/~jcdverha/scijokes/1_2.html

Take a look at these websites. I was surprised to find so many very funny things on math, science, and statistics.

Lastly, I would like to thank my wife, Theresa, and sons, Zach and Justin, for their understanding and patience while I worked on this book. You make every day a joy.

John H. Kranzler

Essentials
for Statistics

If you are one of the "terrified" for whom this book is intended, the chapters in this section may be particularly helpful. Chapter 1 provides an introduction to the text. Because the nature and content of statistics courses typically differ from that of courses in many fields of study, this chapter offers study tips for students of statistics. Chapter 2 presents some general strategies and techniques for dealing with the uncomfortable feelings that many students experience when taking a course in statistics. One common problem experienced by students in statistics courses is not being able to demonstrate on tests what they have learned because of anxiety. If you think you might be one of these people, this chapter may help. Chapter 3 presents an introduction to SPSS 14.0 for Windows, one of the most widely used statistical software programs and the one used throughout this book. This chapter walks you through the basics of SPSS that you will use throughout this book.

The chapters in Section I are intended to help you get off to a running start and may well be worth your time and energy. Of course, if you are already comfortable with numbers and know all the basics of SPSS, you may not get much out of these chapters. Nonetheless, "Heck, I already know all this stuff," is a great way to begin a statistics class. Especially if you think it might be terrifying!

1

Effective Strategies for Studying Statistics

- Effective Strategies for Studying Statistics

> "You haven't told me yet," said Lady Nuttal, "what it is your fiancé does for a living."
>
> "He's a statistician," replied Lamia, with an annoying sense of being on the defensive.
>
> Lady Nuttal was obviously taken aback. It had not occurred to her that statisticians entered into normal social relationships. The species, she would have surmised, was perpetuated in some collateral manner, like mules.
>
> "But Aunt Sara, it's a very interesting profession," said Lamia warmly. "I don't doubt it," said her aunt, who obviously doubted it very much. "To express anything important in mere figures is so plainly impossible that there must be endless scope for well-paid advice on how to do it. But don't you think that life with a statistician would be rather, shall we say, humdrum?"
>
> Lamia was silent. She felt reluctant to discuss the surprising depth of emotional possibility which she had discovered below Edward's numerical veneer.
>
> "It's not the figures themselves," she said finally, "it's what you do with them that matters."
>
> —K. A. C. Manderville, *The Undoing of Lamia Gurdleneck*

Another statistics book! There are now so many statistics books on the market that it seems strange even to me that there can be another one. As someone who has taken statistics courses, worked as a teaching assistant in statistics courses, and taught statistics, I have been dissatisfied with the available books, because they seem aimed at students who whizzed right through college algebra and considered majoring in math just for the sheer joy of it. A number of my students in social science programs are not like that. Many of them would respond with a hearty "true" to many of the following self-test statements. I invite you to test yourself, to see if you too fit the pattern.

1. I have never been very good at math.
2. When my teacher tried to teach me long division in the fourth grade, I seriously considered dropping out of school.
3. When we got to extracting square roots, thoughts of suicide flashed through my mind.
4. Word problems! My head felt like a solid block of wood when I was asked to solve problems like, "If it takes Mr. Jones 3 hours to mow a lawn and Mr. Smith 2 hours to mow the same lawn, how long will it take if they mow it together?"
5. Although I never dropped out of school, I became a quantitative dropout soon after my first algebra course.
6. I avoided courses like chemistry and physics because they required math.
7. I decided early on that there were some careers I could not pursue because I was poor in math.
8. When I take a test that includes math problems, I get so upset that my mind goes blank and I forget all the material I studied.
9. Sometimes I wonder if I am a little stupid.
10. I feel nervous just thinking about taking a statistics course.

Did you answer "true" to some of these items? If so, this book may be helpful to you. When writing it, I also made some assumptions about you:

1. You are studying statistics only because it is a requirement in your major area of study.
2. You are terrified (or at least somewhat anxious) about math and are not sure that you can pass a course in statistics.
3. It has been a long time since you studied math, and what little you knew then has been long forgotten.
4. With a little instruction, and a lot of hard work, you can learn statistics. If you can stay calm while baking a cake or reading your bank statement, there is hope for you.
5. You may never learn to love statistics, but you can change your statistics self-concept. When you finish your statistics course you will be able to say, truthfully, "I am the kind of person who can learn statistics!"

The aim of this book is to help you achieve two important objectives. The first is to deal with math anxiety and avoidance responses that interfere with learning statistics. The second is to understand and compute basic statistics using SPSS, one of the most widely used statistical software programs.

EFFECTIVE STRATEGIES FOR STUDYING STATISTICS

The following is some advice on studying statistics that you will find useful as we move toward these two objectives.

Develop a Solid Math Foundation

Being terrified of math didn't just happen to you overnight. Chances are you have been having bad experiences with math for many years. Most people who have these sorts of bad experiences have not mastered some of the basic rules for working with numbers. Because they don't know the rules, the problems don't make sense. It's sort of like trying to play chess without knowing how the pieces can be moved or what checkmate means. When the problems don't make sense, but everyone else seems to understand them, we are likely to decide that there's something wrong with us. We'll just never be able to do it and, besides, we hate math anyhow. So we tune out, turn off—and a bad situation gets worse.

To make matters worse, most statistics courses are cumulative. New concepts are constantly added to and built upon previous concepts. An introductory statistics course begins with relatively straightforward and concrete concepts, but quickly becomes more complex and abstract. It's important to master the basic math concepts used in virtually all statistics, because that will make understanding the results of statistical analyses much easier.

With the widespread use of statistics programs like SPSS, the days of computing statistics by hand are gone; however, you still need to understand where the numbers come from and what they mean. Appendix A is designed to help get by that kind of thinking by presenting a brief review of basic math concepts that underlie most statistics.

Form a Study Group

Study groups can be extremely helpful while preparing for exams, completing homework, taking notes, sharing learning strategies, and providing emotional support. The optimum size for study groups is three people. Effective groups tend to have no more than five members, however. Exchange telephone numbers and times when you can be reached. Meet regularly in a distraction-free environment. Talk about what you are studying and offer to help others (you may learn best by teaching others). When you are stuck with a problem that you can't solve, don't hesitate to ask others for their help. Very likely they will have some of the same feelings and difficulties you do. Not everyone gets stuck on the same topics, so you can help each other.

Keep Up

If you are in a statistics class, go to class every day and take complete notes. Complete all the assigned reading and homework as soon as possible and before new material is covered. This ensures that new concepts are fresh in your mind and linked to previously learned material. Students who are "terrified" of statistics are susceptible to falling behind, often because of their general dislike of the content. Playing "catch up" in a statistics class is very difficult. Don't let this happen to you. The material in the next chapter on overcoming math anxiety might be helpful if you are one of these people.

http://www.glenbaxter.com/

Time Management

A widely used rule of thumb for the amount of time to spend studying for a college course is 2 hours of study time per credit hour per week. For a 3-credit class, you should plan to spend 6 hours studying outside class each week. Is this enough time for a statistics class for you? Maybe. Maybe not. For many of my students, statistics is the hardest class they will take. You should take as much time as you need to do all the assigned homework and reading and to understand the material. Regardless of the amount of time you need to learn statistics, spread the time you spend studying over a number of days rather than one or two days per week. For example, if you plan to devote 10 hours per week to the study of statistics, spend 2 hours studying at the same time for five days each week. Don't cram 10 hours of study time into one or two days each week!

Study Actively

Be actively involved in the learning process. Take responsibility for understanding the material. Recognize when you don't understand something and get help. Be an active participant in class. But if you're not comfortable asking or answering questions in class, when you have difficulty grasping a new statistical concept seek out assistance from your instructor during office hours, from fellow students in your study group, or from support services or materials (e.g., resource center, peer tutors, professional tutors). This is very important when learning statistics with SPSS, because the computation of statistics is simple. However, you still have to understand why you did what you did and what your results mean.

Practice, Practice, Practice

Statistics is learned best by doing, not by reading a textbook. Don't waste your time memorizing formulas or the steps needed to conduct an analysis in SPSS! Focus instead on the concepts underlying the use and interpretation of statistics. Do the assigned homework and as many other problems and exercises as possible. In the beginning, statistics problems will be straightforward and involve only one step. Increasingly, you will be presented with problems that require you to take several steps to solve them and to make important choices at each step. Keep in mind that these problems can be broken down into a series of small steps, each of which can be solved individually before moving on to the next. Divide and conquer!

2

Overcoming Math Anxiety

- What Causes Math Anxiety?
- Overview of Rational-Emotive Therapy
- Irrational Beliefs
- How to Deal with Math Anxiety

> A statistics major was completely hungover the day of his final exam. It was a true/false test, so he decided to flip a coin for the answers. The statistics professor watched the student the entire two hours as he was flipping the coin... writing the answer... flipping the coin... writing the answer. At the end of the 2 hours, everyone else had left the final except for the one student. The professor walks up to his desk and interrupts the student, saying, "Listen, I have seen that you did not study for this statistics test, you didn't even open the exam. If you are just flipping a coin for your answer, what is taking you so long?" The student replies bitterly (as he is still flipping the coin), "Shhh! I'm checking my answers!"

If you are what might be termed a "math-anxious" or "math-avoidant" person, this chapter may be helpful to you. Most of the material in this chapter is drawn from the theory and practice of rational-emotive therapy (RET), originally developed by the psychologist Albert Ellis. RET has been shown through research to be quite effective in helping people overcome problems like yours. Unfortunately, in a book devoted to statistics, I can only introduce you to some of the basic ideas and techniques. If you are interested, you can enrich your understanding by reading books like Ellis and Harper's *A Guide to Rational Living* or G. D. Kranzler's *You Can Change How You Feel*. (Notice the sneaky way of getting in a plug for my dad's book?)

WHAT CAUSES MATH ANXIETY?

Fear of math, or math anxiety, is what is called a *debilitative emotion*. Debilitative emotions such as math anxiety are problem emotions because (a) they are extremely unpleasant, and (b) they tend to lead to self-defeating behavior, such as "freezing" on a test or avoiding courses or occupations that you otherwise would enjoy.

What you do about your math anxiety (or any other problem) will depend on your theory of what is causing the problem. For example, some people believe that the cause is hereditary: "I get my fear of math from Mother, who always had the same problem." Others believe that the cause lies in the environment: "Women are taught from a very young age that they are not supposed to be good in math, to avoid it, and to be afraid of it." The implication of these theories is that if the cause is hereditary, you can't do much about the problem (you can't change your genetic structure), or if the cause is the culture in which you live, by the time you can change what society does to its young, it will still be too late to help you. Although there may be some truth in both the hereditarian and environmental theories, I believe that they can, at most, set only general limits to your performance. Within these limits, your performance can fluctuate considerably. Though you have very little power to change society and no ability to change the genes you inherited, you still have enormous power to change yourself if you choose to do so, if you know how to bring about that change, and if you work hard at it.

OVERVIEW OF RATIONAL-EMOTIVE THERAPY

Let's begin with the ABC's. A stands for Activating event or experience, such as taking a difficult math test; C stands for the emotional Consequence, such as extreme nervousness. Most people seem to believe that A causes C. In fact, this theory seems to be built right into our language. Consider the following examples:

Activating Event	Cause	Emotional Consequence
(Something happens . . .	that causes me . . .	to feel. . .)
(When you talk about math . . .	that causes me . . .	to feel. . .)
"This test. . .	makes me . . .	nervous."

The implications of this A-causes-C theory are (a) you can't help how you feel, and (b) the way to deal with the problem is to avoid or escape from activating events such as math tests. But is the A-causes-C theory true? Respond to the following items by indicating how you would feel if you were to experience the event. Use a scale that ranges from −5, indicating very unpleasant emotions (such as rage, depression, or extreme anxiety), to +5, indicating an emotion that is extremely positive (such as elation or ecstasy); or use a 0 if you would experience neutral, neither positive nor negative, feelings:

1. Handling snakes.
2. Giving a speech in front of one of your classes.

3. Seeing your 8-year-old son playing with dolls.
4. The death of a loved one in an automobile accident.

I have administered items like these to hundreds of people and have found that for items 1 through 3 the responses have ranged all the way from –5 to +5. On the item concerning the death of a loved one, most people respond with a –5, but when questioned, they have heard of cultures where even death is considered to be a positive event (in the United States everyone wants to go to Heaven but nobody wants to die). Why is it that, given the same Activating event, people's emotional Consequences vary so much?

Differing responses like this suggest that maybe $A \rightarrow C$ isn't the whole story. There must be something else, something that accounts for the different ways people respond to the same stimulus. I believe that it is not A, the Activating event, that causes C, the emotional Consequence. Rather, it is B, your Belief about A, that causes you to feel as you do at point C. Take the example of observing your 8-year-old son playing with dolls. What does a person who experiences feelings of joy believe about what he or she sees? Perhaps something like, "Isn't that wonderful! He's learning nurturing attitudes and tenderness. I really like that!" But the person who experiences very negative feelings probably is thinking, "Isn't that awful! If he keeps that up, he'll surely turn into an effeminate man, or even be gay, and that really would be terrible!"

IRRATIONAL BELIEFS

Ellis has identified some specific beliefs that most of us have learned and that cause us a great deal of difficulty. He calls these beliefs "irrational beliefs." A number of these beliefs have particular relevance to the phenomenon of math anxiety:

- I must be competent and adequate in all possible respects if I am to consider myself to be a worthwhile person. (If I'm not good at math, I'm not a very smart person.)
- It's catastrophic when things are not the way I'd like them to be. (It's terrible and awful to have trouble with statistics.)
- When something seems dangerous or about to go wrong, I must constantly worry about it. (I can't control my worrying and fretting about statistics.)
- My unhappiness is externally caused. I can't help feeling and acting as I do and I can't change my feelings or actions. (Having to do math simply makes me feel awful; that's just what it does to me.)
- Given my childhood experiences and the past I have had, I can't help being as I am today and I'll remain this way indefinitely. (I'll never change; that's just how I am.)

- I can't settle for less than the right or perfect solution to my problems. (Since I can't be a math whiz, there's no sense in trying to do math at all.)
- It is better for me to avoid life's frustrations and difficulties than to deal with them. (Since math always makes me feel bad, the only sensible thing to do is to avoid math.)

Do any of these sound familiar? If they do, chances are good not only that you learned to believe them a long time ago but also that you keep the belief going by means of self-talk. The first step in changing is to increase your awareness of the kind of self-talk that you do. When you think, you think with words, sentences, and images. If you pay attention to these cognitive events, you may notice one or more of the following types of self-talk, which may indicate your underlying irrational beliefs.

Math phobic's nightmare

Catastrophizing

This type of self-talk is characterized by the use of terms or phrases such as "It's awful!" "It's terrible!" or "I can't stand it!" Now, there are some events that most of us would agree are extremely bad, such as bombing innocent people and earthquakes that kill thousands. Chances are good that you will never be the victim of such an event. But your mind is powerful: If you believe that your misfortunes are catastrophes, then you will feel accordingly. Telling yourself how catastrophic it is to do badly on a statistics test will almost guarantee that you will feel awful about it. And that emotional response, in turn, can affect how you deal with the situation. It is appropriate to be concerned about doing well on a test, because concern motivates you to prepare and to do your best. But when you are overconcerned, you can make yourself so nervous that your performance goes down instead of up.

Do you see how all this relates to the first irrational belief on our list? Performing poorly on a statistics test would be awful, because you believe that you must be competent in all possible respects. If you were to fail at something important to you, that would make you a failure: someone who couldn't respect himself or herself. One of the oddest things about irrational beliefs like this is the uneven way we apply them. Your friend could bomb a test, and you'd still think him or her a worthwhile person. But do badly yourself, and the sky falls in!

When you indoctrinate yourself with catastrophic ideas, when you tell yourself over and over again how horrible it would be if you were to perform poorly, then you defeat yourself, because you become so anxious that you help bring about the very thing you're afraid of, or you avoid the experience that could benefit you.

Overgeneralizing Self-Talk

When you overgeneralize, you take a bit of evidence and draw conclusions that go beyond the data. If you experienced difficulty with math as a child, you may have concluded, "I'll never be good at math" or "I'm stupid in math." If you failed a math course, then you tended to think of yourself as a failure who will never be able to succeed, and trying harder would be completely useless.

Rationally, though, failing once doesn't make you a "failure." Because you had difficulty in the past doesn't mean that you will never succeed. If it did, nobody would ever learn to walk! The most pernicious form of overgeneralizing is self-evaluation. We have a tendency to tie up our feelings of self-worth with our performance. When we do well at something, we say, "Hey! I'm a pretty good [or competent or worthwhile] person!" But when we perform poorly, we tend to believe that we are now worthless as a person. This process begins in childhood. When Johnny does something we consider bad, we tend to encourage overgeneralization by saying, "Johnny, you are a bad boy" (i.e., you are worthless as a person).

If you were a worthless or stupid person, you wouldn't have gotten far enough in your education to be reading this book. True, in the past, you may have had difficulty in math, and math may be difficult for you now. But how does that prove you can't learn it? There is absolutely no evidence that your situation is hopeless or that it is useless to try. The only way to make it hopeless is to tell yourself, over and over, how hopeless it is.

Demanding Self-talk

This type of self-talk includes the use of words such as *should, must,* and *need.* If you are math-anxious, chances are that you use these words to beat up on yourself. You make a mistake and say, "I shouldn't have made that mistake! How could I have been so stupid?" I have a tennis partner who informed me that she finds it difficult to concentrate on her work for the rest of the day after she has played poorly. She believes that she should have done better. Instead of being calmly regretful for having made some errors and thinking about how to do better next time, she bashes herself over the head psychologically for not doing perfectly well, every time.

"But," you may say, "I need to be successful" or "I have to pass this course." Have to? The first time? Or you can't survive? It would be nice to be successful given the advantages it would bring you, but lots of people do manage to function in life even after doing badly in a statistics course. To the degree that you believe you need a certain level of performance, to that degree you will experience anxiety about possible failure and thereby increase the chance of failure.

HOW TO DEAL WITH MATH ANXIETY

What can you do about a way of thinking that seems so automatic, so ingrained? Here is a series of steps that will probably help. I'd suggest that you try them out, in order, even though you may not expect them to work for you. You might just be surprised!

Step 1. Record your feelings (C)

When you notice that you are feeling anxious, guilty, angry, or depressed about some aspect of your statistics course, record your emotional experience. Describe your feelings as accurately as you can. You might write things such as, "I feel guilty about not having taken more math as an underclassman," or "I feel really nervous about the test we're having next week," or "I'm too shy to ask questions in class," or "I just get furious that they make us take statistics when I'll never have to use it." Write down all the unpleasant feelings you have at the time. When you have done this, you will have described C, the emotional Consequence part of the ABC paradigm.

Step 2. Describe the activating event or experience (A)

Briefly write down what it was that seemed to trigger your feelings. Here are some common activating events for math anxiety. When you write your own, record the thing that is most likely to have happened. Find the immediate trigger, the thing that happened just before your experiencing of the negative emotion. Here are some examples:

- I was assigned some difficult statistics problems, and I don't know how to do them.
- I thought about a test coming up, one that I will almost surely fail.
- I discovered that I need more information about some of the material, but I'm afraid to ask about it in class because I'll look stupid.

Step 3. Identify your irrational beliefs (B)

As accurately as you can, record what you were saying to yourself before and during the time when you experienced the emotions you recorded in Step 1. The first few times you do this, you may have difficulty, because you don't usually pay much attention to the thoughts that seem to race through your head. Although it is difficult to become aware of your thoughts, it is not impossible. One technique you can use is to ask yourself, "What must I have been saying to myself about A (the activating event) at point B in order to experience C (the emotional consequence)?" Suppose your first three steps looked like this:

Step 1. (Describing C, the emotional Consequence) I feel really nervous and miserable.

Step 2. (The Activating event, A) My advisor told me I need to take a statistics class.

Step 3. Identify B, the Belief that leads from A to C. Obviously, you're not saying, "Wow, I'm really going to enjoy that class!" You must have been saying something like:

"If I fail, that'll be awful!"

"I'll be a real loser!"

"I'll never be any good at statistics!"

"I'm going to have a terrible term and hate every minute of it."

"What will the other students and the professor think of me when I do badly?"

Step 4. Challenge each of the beliefs you have identified

After you have written down your self-talk in Step 3, look at each statement and dispute it. One question you can ask to test the rationality of any belief is, "Where's the evidence for this belief?" Let's look at each of the examples listed in Step 3 above:

1. Where's the evidence that it will be awful if I fail? True, failure would be unfortunate, but would it be catastrophic? I'd do better to remember that if I'm overconcerned with doing well, I will be even more likely to fail.

2. Where's the evidence that if I fail the test, I, as a person, will be a failure? The worst I can possibly be is an FHB (a fallible human being) along with the rest of the human race.

3. Where's the evidence that I'll never be good in statistics? I may have some evidence that similar material was difficult for me in the past, but how can that prove anything about the future?

4. Where's the evidence that I will hate every single minute of the term? This statement has several irrational beliefs to be challenged: (a) that I'll hate the course (I might have a great teacher, with a wonderful sense of humor, and actually enjoy it), (b) that the discomfort will generalize to the entire term (I might dislike my statistics class but very much enjoy my other courses), and (c) that I will spend every single minute of the term feeling hateful (no help needed to challenge this one, right?).

5. This statement appears to be a rhetorical question. Chances are I'm not really wondering what others will think of me if I fail, but rather telling myself all the bad things they'll think—and how awful that will be. Both parts of this can be challenged: Where's the evidence that they'll think bad things about me and, even if they do, would that be catastrophic?

Step 5. Once you have identified and challenged an irrational belief, the next step is to replace it with a rational one

Ask yourself what you would rather believe—what your best friend might believe?—what Harrison Ford or Albert Einstein or Desmond Tutu probably would believe? Then, every time you find yourself moving into that old irrational self-talk, answer it with the new alternative belief.

Step 6. Do rational-emotive imagery

After you have practiced replacing your irrational beliefs a few times, you may feel better. Some people, however, report that they now understand that their beliefs cause their unpleasant emotions, and they realize that those beliefs are irrational, but they still feel much the same way as before. If this is true of you, you may benefit from doing some imagery. I will discuss both mastery and coping imagery techniques, because some of my students have reported that one approach is more effective for them than the other. Before attempting either kind of imagery, however, do spend several days practicing Steps 1 through 5.

Mastery Imagery.   In the mastery imagery approach, you are to imagine yourself mastering the situation, that is, feeling and acting in an appropriate way in the presence of the activating event. If you are anxious about a statistics test, imagine feeling calm or at most only slightly concerned while taking the test, answering the questions as well as you can, calmly leaving the exam, and being able to look back on the experience with some satisfaction. Imagine speaking rationally to yourself during the whole experience (taken from your material in Step 5). Make the image (the daydream, if you want to call it that) as vivid and real as possible. If you feel very upset during the experience, terminate the imagery; go back and reread Step 4 and attempt the imagery again the next day. For any kind of positive imagery to be effective, you will need to work at it for at least a half-hour per day for a week; don't expect immediate results.

Coping Imagery.   Again, imagine yourself in the experience that you're having problems with, for example, taking a statistics test. This time include having difficulty and starting to experience anxiety. Then imagine dealing with the anxiety by saying to yourself, "Stop! Relax!" Try to force yourself to feel more calm. Breathe deeply a few times, remind yourself of rational self-talk, and try to change the extremely anxious feelings to ones that are more calm. Imagine coping with the problem. Again, you won't experience immediate success; it usually takes at least a week of imagery work before you begin to get results.

Step 7. Activity homework

You can only live in your imagination so long if you want to attain objectives in the real world. Sooner or later you need to take a deep breath and *do something*. If you experience math anxiety, one such "something" might be to work your way through this book. As you begin to make these sort of conscious, real-world changes, be aware of your self-talk. When you notice yourself feeling anxious or emotionally upset, dispute your irrational beliefs as actively as you can. If things don't get better immediately, don't give up—keep using these techniques for at least a couple of weeks. Remember, the odds are in your favor!

3

How to Use SPSS

- What is SPSS?
- Starting SPSS
- Basic Steps in SPSS Data Analysis
- Finding Help in SPSS

> Three men are in a hot-air balloon. Soon, they find themselves lost in a canyon somewhere. One of the three men says, "I've got an idea. We can call for help in this canyon and the echo will carry our voices far." So he leans over the basket and yells out, "Hellllooooooo! Where are we?" (They hear the echo several times.) 15 minutes later, they hear this echoing voice: "Hellllooooooo! You're lost!!" One of the men says, "That must have been a statistician." Puzzled, one of the other men asks, "Why do you say that?" The reply: "For three reasons: (1) he took a long time to answer, (2) he was absolutely correct, and (3) his answer was absolutely useless."

Today's software programs make life much easier for students who are learning statistics. Gone are the days of entering data on a hand-held calculator and following the steps of a complex formula to calculate statistics. Most of the work in programs like SPSS is done by simply pointing and clicking. In SPSS, you can enter data, run analyses, and display your results in tables and graphs in a matter of minutes. Sound great? It is.

WHAT IS SPSS?

The purpose of this chapter is to introduce you to the basics of SPSS 14.0 for Windows. Throughout this book, whenever we discuss something in statistics, I'll show you how to do it in SPSS at the same time. Please keep in mind that SPSS is a very powerful program. Because this book covers basic concepts in statistics, we will only cover what we need to do in SPSS to meet that objective. There are many things in SPSS that we will not discuss that are beyond the scope of this book.

"I THINK YOU SHOULD BE MORE
EXPLICIT HERE IN STEP TWO."

SidneyHarris/www.sciencecartoonsplus.com

After reading this chapter, if you find you want more information on how to use SPSS, I encourage you to review the online tutorial that comes with the program. The tutorial is designed to familiarize readers with many of the features of SPSS. You may also want to obtain the SPSS *Brief Guide* as a supplement to the online guide.

STARTING SPSS

To start SPSS, click on the name of the program in the Start Menu, just like you would for any other Windows program. After it starts, the first dialog box you are presented with in SPSS is shown in Figure 3-1.

The opening dialog box asks you "What you would like to do?" and presents you with a number of choices, ranging from running the online tutorial to opening an existing data source. You can check the box in the lower left corner titled "Don't show this dialog in the future" if you don't want to see this box

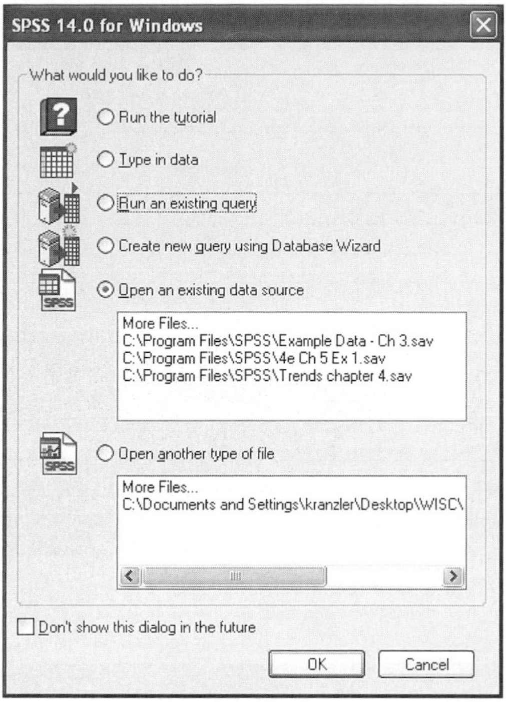

Fig 3-1 The SPSS opening dialog box.

again. For now, click *Cancel* in the lower right-hand corner of the box. After you do this, the *Data Editor* opens automatically with the title *Untitled—SPSS Data Editor* at the top of the window (see Figure 3-2).

Just under the title is the menu bar. Most of the things you need to do in SPSS start by selecting an option in one of the menus. Each window in SPSS has its own menu bar with menus appropriate for that window. Rather than provide an overview of all the various menu options, we'll discuss them as needed. The *Data Editor* is used to create new data files or to edit existing ones. The Data Editor actually consists of two windows called *Data View* and *Variable View*. The Data View window is shown here. You can move between the Data View and Variable View windows by clicking on the tabs in the lower left-hand corner of the Data Editor. Click on the Variable View tab to see the Variable View window, as shown in Figure 3-3.

Both windows in the Data Editor display aspects of the contents of a data file. In Data View, the rows represent cases (e.g., the scores for each participant in a research study) and the columns represent variables (i.e., the things we want to measure); and in Variable View, the rows represent variables and the columns represent the attributes of each variable. In other words, Data

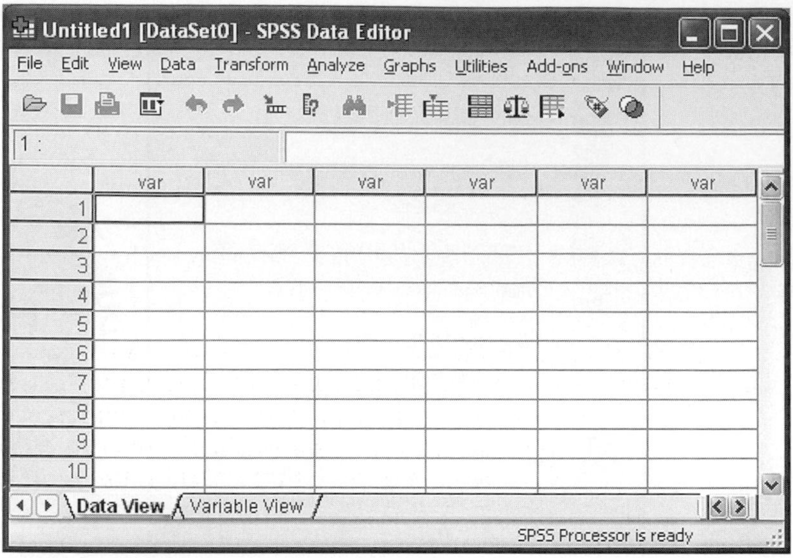

Fig 3-2 SPSS data editor (data view).

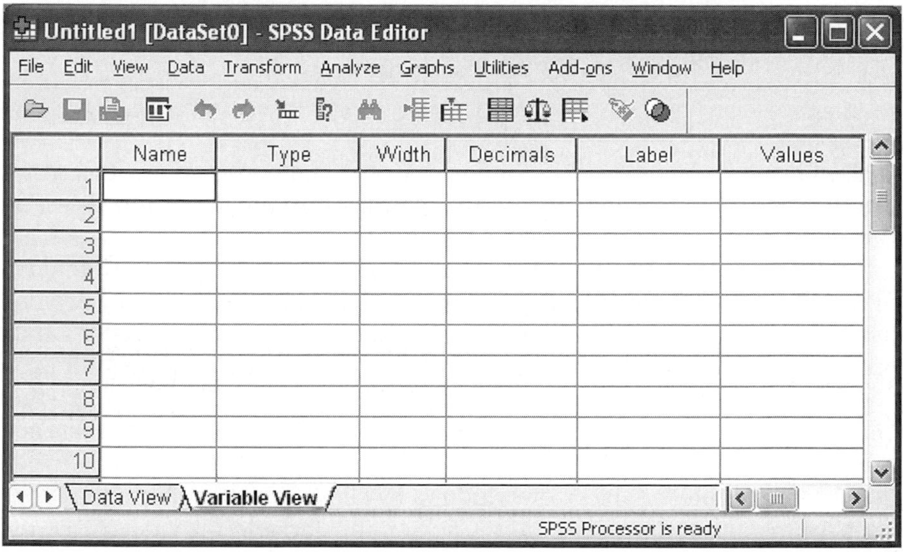

Fig 3-3 SPSS data editor (variable view).

View contains the data that will be analyzed and Variable View contains information about the variables themselves. We'll talk more about different kinds of variables in Chapter 4.

Now that you are familiar with the Data Editor, let's enter some data and run a simple analysis so that you have an overview of the main components of SPSS that we will use throughout the book.

BASIC STEPS IN SPSS DATA ANALYSIS

Analyzing data in SPSS is very straightforward. Almost everything you need to do can be done in four basic steps:

1. *Enter data.* When you first open SPSS, the Data Editor is ready for data to be entered. You can get data into the Data Editor in several ways. You can enter it directly by typing it in or you can import data from another program (e.g., such as a spreadsheet). You can also open files that already have been entered into the Data Editor and saved in SPSS.
2. *Select an analysis.* After you have entered your data, you can select a procedure from one of the menus at the top of the Data Editor. We will be mainly using the *Analyze* and *Graphs* menus.
3. *Select variables to analyze.* After you select a procedure in SPSS, you are presented with a *dialog box* so that you can select the variables you want to analyze.
4. *Run analysis and examine results.* To run an analysis in SPSS, you simply point and click. After you have run an analysis, the *Viewer* window opens and displays the results.

That's it! Now let's walk through an example so this is clear.

Entering Data

Before you enter data into the Data Editor, you need to define the variables themselves. To define variables, click the tab for Variable View at the bottom of the Data Editor window. In Data View, each row is a different variable. As an example, let's say you are interested in the relationship between reading comprehension and gender in school-aged children. For your study, you obtain the following information for a small group of children: age, gender, and score on a standardized test of reading comprehension. To define these variables in Variable View:

- In the first row of the Name column, type Age and press the Enter key;
- In the second row of the Name column, type Reading and press the Enter key;
- In the third row of the Name column, type Gender and press the Enter key.

You have just created and labeled your three variables. If you make a mistake, just backspace and retype the variable name. Variables can be given any name, provided that they have no more than eight characters. You may have noticed that each time you pressed the Enter key, SPSS automatically defined the variable type as *Numeric*. Numeric variables are variables with numbers only, such as age and reading test score. You can also enter nonnumeric data, such as strings of text, into the Data Editor. Gender, in our example, is a nonnumerical—or string—variable. To change the variable type for Gender, click on the cell for Gender in the third row of the *Type* column, as shown in Figure 3-4.

After you click on the . . . button on the right side of the cell, you will be presented with the *Variable Type* dialog box shown in Figure 3-5.

To specify the appropriate type for this variable, select *String* and then click *OK* to save your changes and return to the Variable View window. If you

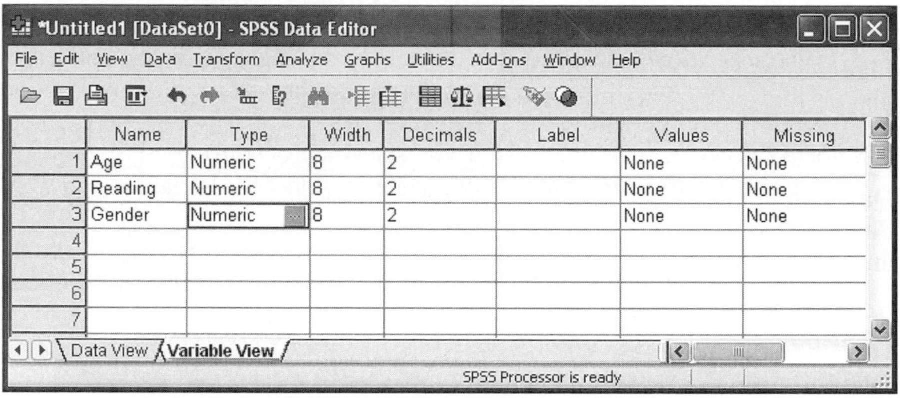

Fig 3-4 Defining variables in variable view.

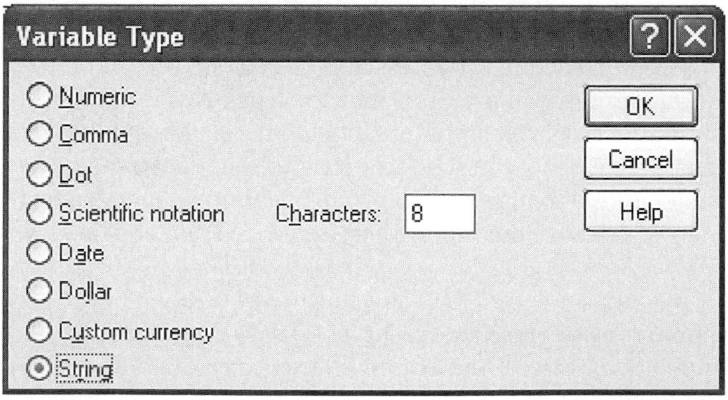

Fig 3-5 Variable type dialog box.

want to change the number of decimals for your variable, click on the cell for the variable you want to change in the *Decimals* column and indicate the number of decimals desired by clicking on the up or down arrow, as shown in Figure 3-6.

For our example, let's work with whole numbers only, so select the number of decimals to be zero for each variable. As you can see, this is the default setting for string variables.

You can also provide more information on variables in the *Label* column. To label the variables in our example:

- In the first row of the Label column, type Student Age in Years and press the Enter key;
- In the second row of the Label column, type Reading Test Score and press the Enter key;
- In the third row of the Label column, type Student Gender and press the Enter key.

Figure 3-7 displays the variables that we just defined. For now, let's not worry about the other columns in Variable View.

Now that we have defined the variables, click on the Data View tab to return to the Data View window to enter our example data. As you can see in Data View, the names of the variables that we created in the Variable View window appear in the first three columns. Let's begin entering data in the first row and the first column.

- In the *Age* column, type "6" and press the Tab key;
- In the *Reading* column, type "100" and press the Tab key;
- In the *Gender* column, type "Male" and press the Tab key.

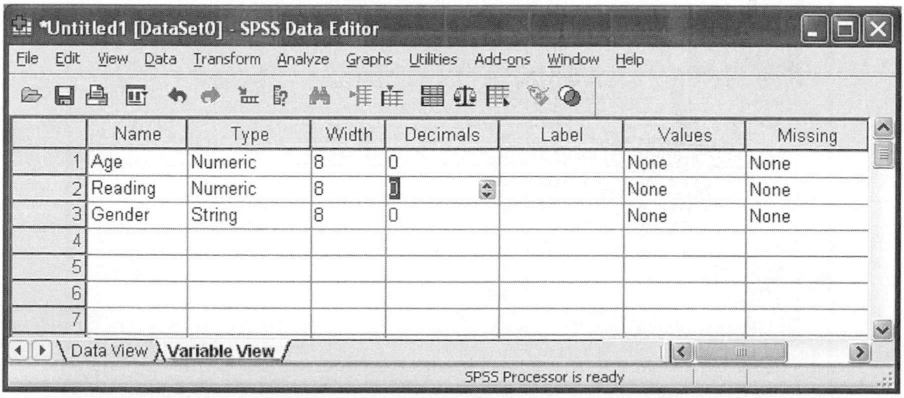

Fig 3-6 Selecting the number of decimals.

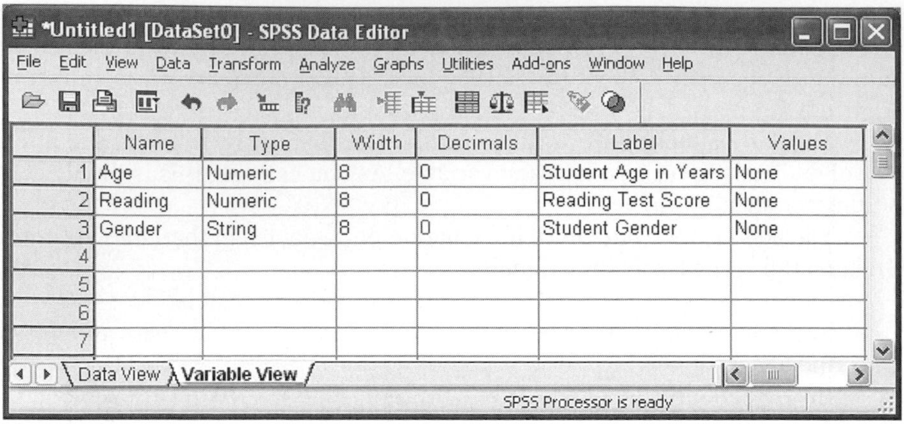

Fig 3-7 Variable view of example variables.

We have now entered the data for one of the participants in the example study. According to our data, this is a 6-year-old boy who obtained a score of 100 on the reading test. Now enter the data for the other four cases to complete the set of data, as shown in Figure 3-8.

Before we move on to the next step, let's save our data in SPSS. To save a file in SPSS, click on *File* in the menu bar at the top of the Data Editor and then on *Save*. When the dialog box appears, type in "Example Data" as show in Figure 3-9 and then click *Save*. This saves a copy of your dataset with the file-name "Example Data."

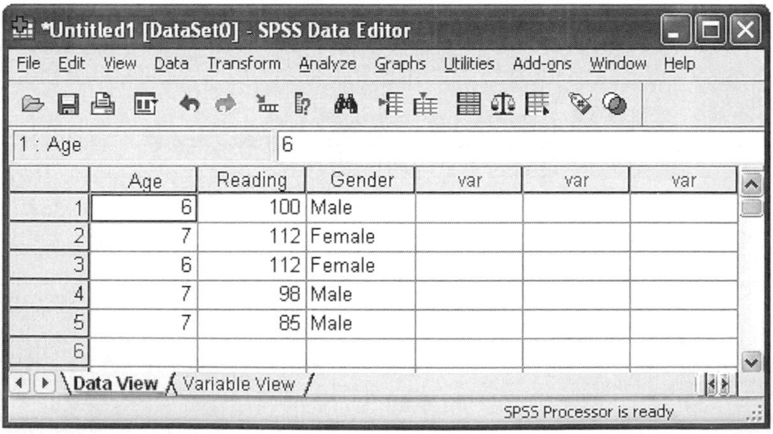

Fig 3-8 Example data in data view.

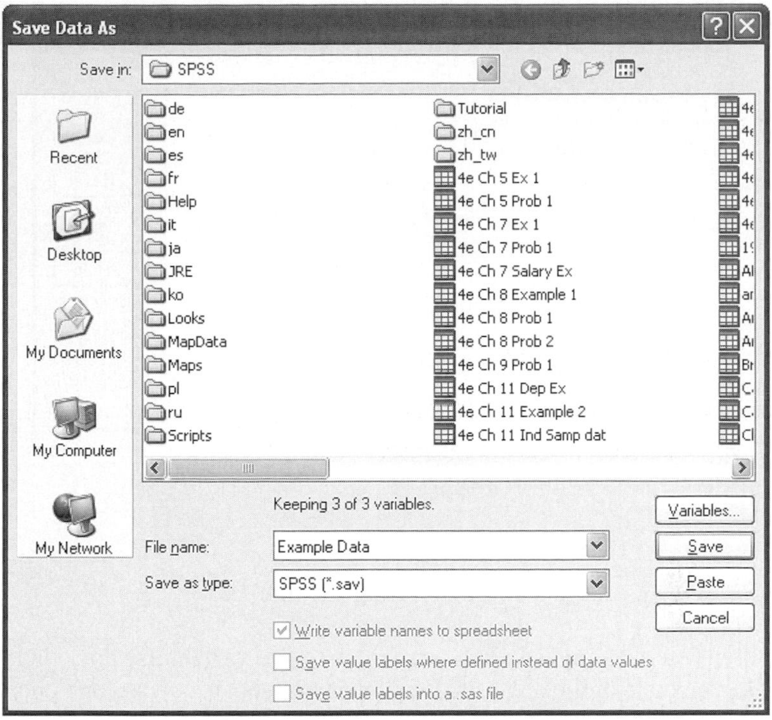

Fig 3-9 Save data as dialog box.

Selecting an Analysis

After you have entered the example data and saved the data file, you are ready to select a procedure from the menus to conduct a statistical analysis or to create a chart. The *Analyze* menu in the Data Editor contains a list of all the analysis groups that can be conducted in SPSS, with several analyses within each grouping. Clicking on an item in a list will display the analyses that can be selected within each group.

For our example study, let's examine the frequency with which each score was obtained for each variable. To do this, move your mouse over the *Descriptive Statistics* item in the Analysis menu and click on *Frequencies*. After you have done this, the *Frequencies* dialog box in Figure 3-10 will appear, displaying a list of the variables in our study.

Now you must select the variables you want to analyze. For our example, let's analyze all the variables. To do this, click on the first variable, *Student Age in Years*, then click on the ▶ in the middle of the dialog box. As you can see, this moves the variable to the *Variable(s)* box. Now do the same thing for the other two variables. Also make sure the *Display frequency tables* box in the lower left-hand corner is checked.

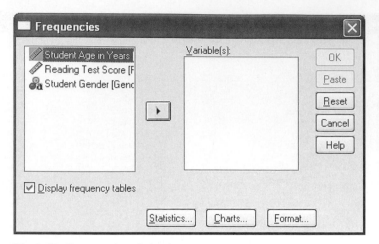

Fig 3-10 Frequencies dialog box.

Running an Analysis and Examining Results

After you have moved all the variables to the Variable(s) box, click *OK* in the dialog box to run the analysis. Results of statistical analyses and charts created in SPSS are displayed in the Viewer window. Results for our example are displayed in Figure 3-11.

As you can see in this figure, the output for this analysis consists of the number and percentage of each score for each of our variables. You've now done your first statistical analysis in SPSS! Congratulations. To save the contents of the Viewer, click on *File* in the menu bar and then *Save*. You then provide a name for the output just like you did when saving the data file. The only difference between the two types of files is the output is saved with the .spo extension. To exit SPSS, click on *File* in the menu bar, then select *Exit*.

FINDING HELP IN SPSS

Getting help in SPSS is simple and easy to do. Every window in SPSS has a *Help* menu on the menu bar to provide you with assistance. The *Topics* menu item in the *Help* menu provides a list of all the topics for which you can get help. In addition, many of the dialog boxes have a *Help* button that will provide you with help specifically for that dialog box. Finally, the *Tutorial* menu item in the *Help* menu provides access to the introductory tutorial. This is a very helpful overview of the SPSS system and covers many things that we will not address in this book.

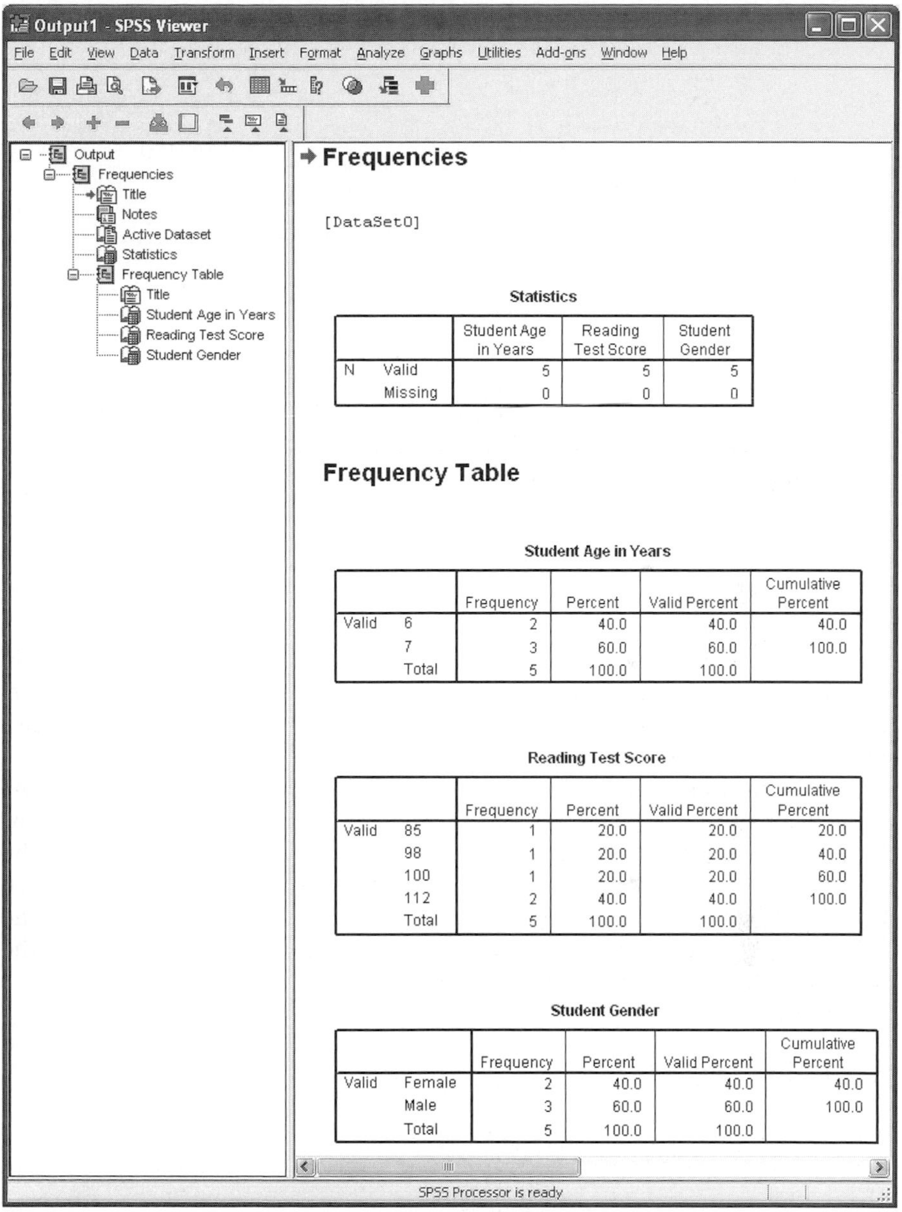

Frequencies

[DataSet0]

Statistics

		Student Age in Years	Reading Test Score	Student Gender
N	Valid	5	5	5
	Missing	0	0	0

Frequency Table

Student Age in Years

		Frequency	Percent	Valid Percent	Cumulative Percent
Valid	6	2	40.0	40.0	40.0
	7	3	60.0	60.0	100.0
	Total	5	100.0	100.0	

Reading Test Score

		Frequency	Percent	Valid Percent	Cumulative Percent
Valid	85	1	20.0	20.0	20.0
	98	1	20.0	20.0	40.0
	100	1	20.0	20.0	60.0
	112	2	40.0	40.0	100.0
	Total	5	100.0	100.0	

Student Gender

		Frequency	Percent	Valid Percent	Cumulative Percent
Valid	Female	2	40.0	40.0	40.0
	Male	3	60.0	60.0	100.0
	Total	5	100.0	100.0	

SPSS Processor is ready

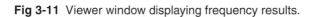

Fig 3-11 Viewer window displaying frequency results.

Describing Univariate Data

Isn't this a great statistics book? Here we are on Chapter 4 and we haven't even begun covering statistics yet! Well, I hate to spoil the fun, but the time has come. If you apply what you learned in Section I, however, you're ready for the challenge. The purpose of this section is to teach you how to describe univariate data—that is, information gathered on a single variable. Chapter 4 will describe some straightforward techniques for summarizing data in the form of frequency distributions and displaying that information graphically. In Chapter 5 you will learn how to calculate and interpret descriptive statistics to summarize the level and variability of data. Chapter 6 presents a discussion of the characteristics of one particularly important frequency distribution for statistics called the *normal distribution*. Finally, Chapter 7 explains how to interpret percentiles and standard scores.

4

....

Frequency Distributions

- What are Statistics?
- Variables
- Scales of Measurement
- Frequency Distributions
- Graphing Data
- Normal Curve
- Problems

Two statisticians were traveling in an airplane from L.A. to New York. About an hour into the flight, the pilot announced that they had lost an engine, but don't worry, there are three left. However, instead of 5 hours it would take 7 hours to get to New York. A little later, he announced that a second engine failed, and they still had two left, but it would take 10 hours to get to New York. Somewhat later, the pilot again came on the intercom and announced that a third engine had died. Never fear, he announced, because the plane could fly on a single engine. However, it would now take 18 hours to get to New York. At this point, one statistician turned to the other and said, "Gee, I hope we don't lose that last engine, or we'll be up here forever!"

WHAT ARE STATISTICS?

What are statistics? Statistics are a broad range of techniques and procedures for gathering, organizing, analyzing, and displaying quantitative data. "Data" means information: Any collection of information is a collection of data. For statisticians, "data" generally refers to quantitative information, that is, something that can be expressed in numbers (e.g., quantity or amount). There are two main kinds of statistics: descriptive and inferential. *Descriptive statistics* are used to describe a set of quantitative data. *Inferential statistics* are used to make inferences about large groups of people (i.e., populations) by analyzing data gathered on a smaller subset of the larger group (i.e., samples). Results of these analyses are used to make inferences about the larger group.

VARIABLES

For those of us interested in the social sciences (e.g., education, psychology), the data we gather often involve the characteristics of people. When observing people, what's one of the first things you notice? That's right—we differ, sometimes a little and sometimes a lot. In fact, people differ on virtually every biological and psychological characteristic that can be measured. Familiar examples of physical characteristics on which people differ include height and weight, blood type and pressure, body temperature, visual acuity, and eye color. But we also differ on intelligence, academic achievement, temperament, personality, values, and interests, among many other traits. In statistics, characteristics on which people differ are called *variables*. Variables can be either *discrete* or *continuous*. Discrete variables can only take on certain values. For example, the

number of people in a household is a discrete variable. You can have 1, 2, 3, or 4 people in a household, for example, but not 1.5. Continuous variables are characteristics that can take on any value (e.g., height, weight).

In research, variables can also be categorized as *independent* or *dependent*. An independent variable is a variable that is manipulated to determine its effect on another variable. A dependent variable is the focus of most statistical analyses, because it is the variable that is measured in response to manipulation of the independent variable.

SCALES OF MEASUREMENT

Measurement refers to the assignment of numbers to the characteristics on which people differ (variables). Different kinds of variables require different rules for assigning numbers that accurately reflect how people differ on those variables. Not all variables can be assigned numbers according to the same rules. It depends on what you are trying to measure. Variables can be measured on one of four different *scales* of measurement: nominal, ordinal, interval, and ratio. Each scale has a particular set of rules that defines how numbers are assigned to variables and what you can do with those numbers in terms of statistics.

Nominal Scale.   Variables that are measured on a nominal scale are often referred to as qualitative or categorical variables. Measurement on a nominal scale involves the assignment of people or objects to categories that describe the ways in which they differ on a variable. Examples include gender (Male, Female), marital status (Single, Married), eye color (Blue, Green, Brown), and race/ethnicity (Caucasian, African American, Hispanic, Asian, Other). All people or objects within the same category are assumed to be equal. On a nominal scale of measurement, numbers are often used to stand for the names or labels of each category (e.g., Male = 1, Female = 2). The number assigned to each category is completely arbitrary, however, and no rank ordering or relative size is implied.

Ordinal Scale.   On an ordinal scale of measurement, it is possible to rank persons or objects according to magnitude. Numbers on this scale are used to rank order persons or objects on a continuum. The continuum used depends on the variable. Variables measured on an ordinal scale include class rank (Frosh–Senior), socioeconomic status (Poor–Rich), and Olympic marathon results (First–Last). For example, on an ordinal scale the first-place finisher in the Olympic marathon would be assigned a rank of 1, the second-place finisher a rank of 2, and so on. On an ordinal scale, these numbers (ranks) express a "greater than" relationship, but they do not indicate "how much greater." Although we know that first place is better than second place, we do not know anything about how close the race was. We don't know whether the top two finishers differed by a tenth of a second, 10 seconds, or 10 minutes; nor can we assume that the difference between first and second place is the same as that between second and third, and so on. This is an important point about ordinal scales of measurement—while the numbers assigned on an ordinal scale do reflect relative merit, the units of measurement are not equal (e.g., $3 - 2 \neq 2 - 1$).

Interval Scale.   On an interval scale, numbers that are assigned to variables reflect relative merit and have equal units of measurement. When a variable has equal units of measurement, it means that the same difference between two points on a scale means the same thing in terms of whatever you are measuring. An example of a variable assumed to be measured on an interval scale is intelligence. On tests of intelligence, we can say that a person with an IQ of 120 is more intelligent than a person with an IQ of 100. We also know that the 20-point difference between IQ scores of 120 and 100 means the same thing in terms of intelligence as the 20-point difference between scores of 90 and 70. Variables measured on an interval scale lack an *absolute zero point*, or the absence of the characteristic being measured, however. For example, there is no absolute zero point of intelligence, or no intelligence at all, although in an election year one is tempted to argue otherwise. Because it is impossible to establish a true zero point, it is not possible to speak meaningfully about the *ratio* between scores. For example, we cannot say that a person with an IQ of 140 is twice as smart as someone with an IQ of 70.

Ratio Scale.   Variables measured on an absolute scale have a true zero point and equal units of measurement. Many physical characteristics such as height, weight, and reaction time are measured on an absolute scale. The true zero point not only indicates the absence of the thing being measured (e.g., no height at all), it also designates where measurement begins. Equal units of measurement provide consistent meaning from one situation to the next and across different parts of the scale. For example, 12 inches in Eugene, Oregon is the same as 12 inches in Gainesville, Florida. Further, the difference between 12 and 24 inches is identical to the difference between 100 and 112 inches. We can also determine ratios on a ratio scale. For example, we can say that 12 inches is half as long as 24 inches, or that it took somebody twice as long to run a marathon as somebody else.

Scales of Measurement in SPSS

In SPSS, you can only define the scale of measurement for variables as nominal, ordinal, and scale. No distinction is made in SPSS between interval and ratio data, because they are treated the same way in statistical procedures. The interpretation of data for these kinds of variables does differ, however, as we noted above. Nominal and ordinal variables in SPSS may consist of either string or numeric data. Scale variables are numeric data that are on an interval or ratio scale.

Specifying the scale of measurement is important for creating charts and tables in SPSS. The scale of measurement for variables is defined in Variable View in the Data Editor. To define the measurement scale for variables, simply

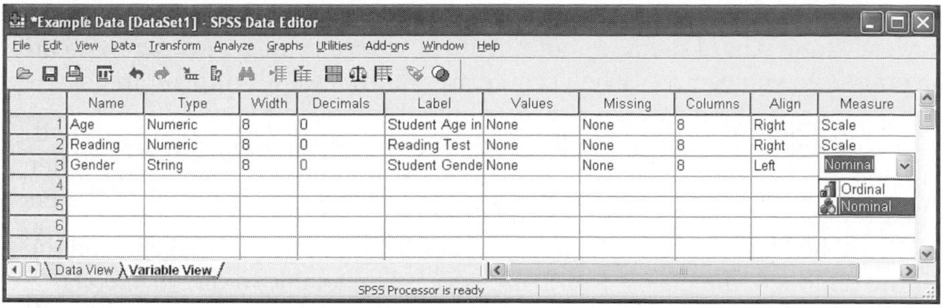

Fig 4-1 Defining the scale of measurement in SPSS.

click on any cell in the *Measure* column, as shown in Figure 4-1, and select the scale of measurement that is appropriate for that variable. As you can see in the figure, string variables can only be defined as ordinal or nominal.

FREQUENCY DISTRIBUTIONS

Whenever you gather data, the initial result is some unordered set of scores. A common first step in the examination of data is to create a frequency distribution. Frequency distributions organize and summarize data by displaying in tabular form how often specific scores were obtained. Imagine that you have administered a test of emotional intelligence to 100 college students. Emotional intelligence refers to one's ability to understand and regulate emotions. On this test, higher scores reflect more emotional intelligence and lower scores reflect less emotional intelligence. The scores they earned are as follows (in no particular order):

				Emotional Intelligence Test Scores								
46	50	48	47	48	47	49	43	47	46	50	48	49
46	46	45	46	46	47	46	46	46	48	47	46	47
44	49	47	48	49	48	48	49	45	48	46	46	51
48	44	45	44	46	49	50	48	43	48	46	48	46
48	46	47	47	47	47	49	49	46	47	47	44	45
45	48	50	48	47	47	49	47	45	48	49	45	47
47	44	48	47	47	51	47	46	47	46	45	47	45
45	47	48	48	46	48	45	50	47				

What can we say about these data? Not much. About all we can say is that most of the scores appear to be in the 40s and a few are in the 50s. Suppose you obtained a score of 48 on this test. What would that tell you? Are you more emotionally intelligent than the average college student or less? Just from perusing this table you can see that a 48 is not the highest score; nor is it the lowest. But it's hard to know more. Are you above average, average, or below average? As you can see, when scores are unordered, it's difficult to get a feel for the data. Making a frequency distribution can help.

To make a frequency distribution in SPSS, click on *Analyze* in the menu bar, then on the *Descriptive Statistics* item, and finally on *Frequencies....* This opens the Frequencies dialog box shown in Figure 4-2.

In the Dialog box, click on *Emotional Intelligence* and then click on the ▶ in the middle of the dialog box. This moves the variable to the *Variable(s)* box so that it will be analyzed. If you have more than one variable, simply repeat this process for all the variables you want to analyze. After you have moved our variable to the Variable(s) box, click *OK* to run the analysis. Figure 4-3 shows the results of the frequency analysis in the Viewer.

For these results, we can see in the box at the top of the window that the total number (*N*) of test scores in our analysis is 100 and that we have no missing data. The box titled *Emotional Intelligence* displays the frequency distribution. The different scores that were obtained are shown in the first column. Here we can see that scores on the test of emotional intelligence ranged from a low of 43 to a high of 51. We can also see the frequency of each score—that is, the number of individuals who obtained each score—in the next column. As is shown, two students obtained a score of 43, five students a score of 44, and so on. The score obtained most often was 47 with a frequency count of 25. The next two columns display the percentage of individuals who obtained each score and the cumulative percent. The cumulative percent can be used to determine the percentage of scores that fall above or below a particular score. Looking at our distribution of scores, we can also see that most of the scores

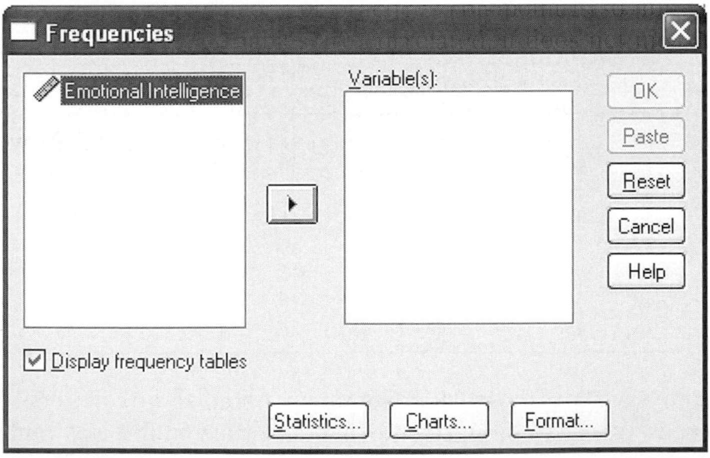

Fig 4-2 Frequencies dialog box.

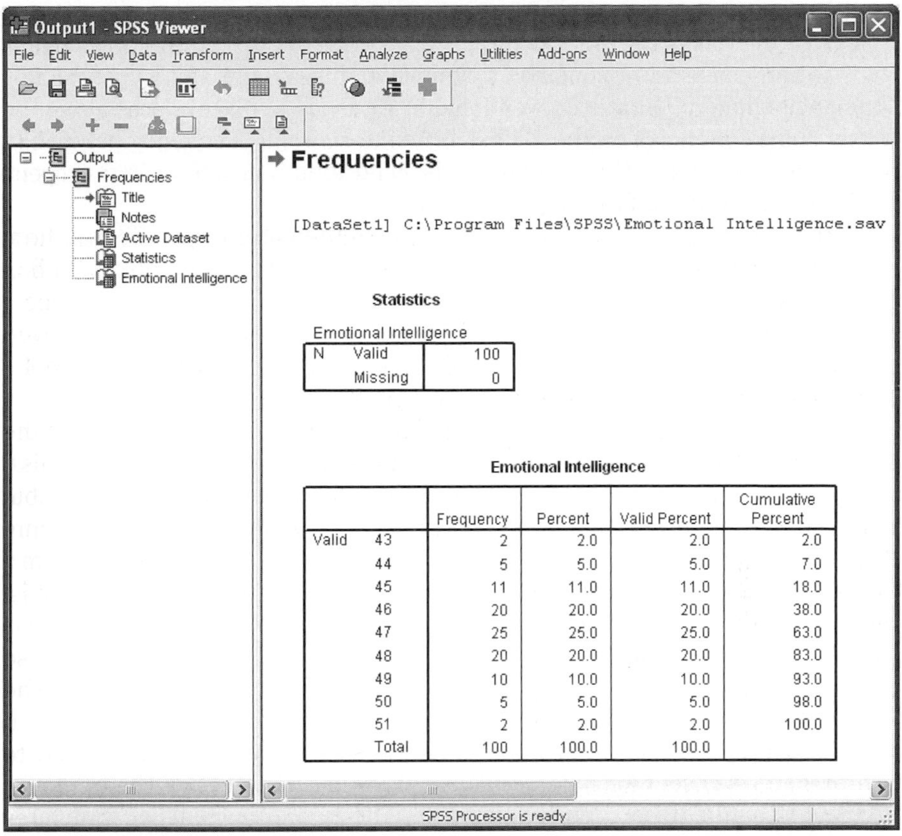

Fig 4-3 Frequency distribution results.

tend to be clustered around the middle of the distribution, with relatively few scores at the extremes. In fact, 65% of our scores are between 46 and 48. Further, we can see that there are no large gaps between scores.

Now what can we say about your hypothetical score of 48? As we can see from the frequency distribution, a score of 48 is slightly above average compared to the other students in the class, surpassing 63% of all other scores. That's a lot more than we knew before.

GRAPHING DATA

Frequency distributions can also be displayed graphically as a frequency polygon (smooth-line curve) or as a histogram (bar graph). Let's talk in some detail about one of the most important kinds of graphs to understand in the study of statistics—the frequency polygon.

Frequency Polygon

Graphs have a horizontal axis (known as the *X*-axis) and a vertical axis (known as the *Y*-axis). It is conventional in statistics to place scores along the *X*-axis and frequencies on the *Y*-axis. To make a simple frequency polygon in SPSS, click on the *Graphs* in the menu bar and then on the *Line* item to open the *Line Charts* dialog box shown in Figure 4-4.

In the dialog box, first click on *Simple* and then on *Define*. This will open the *Define Simple Line: Summaries for Groups of Cases* dialog box shown in Figure 4-5.

In the dialog box, click on the *Emotional Intelligence*, then on the next to the *Category Axis:* area. Also select *N of Cases* in the *Line Represents* area of the dialog box. Now click *OK* to make the line graph shown in Figure 4-6.

This graph displays the frequency distribution of emotional intelligence scores in the form of a line graph. Just in case you haven't worked much with graphs, I'll go through this one slowly. First, look at the humped line that makes up the shape of the graph. Specifically, look at the point of the line directly above the score of 47. Notice that this point is across from the frequency of 25, which indicates that 25 persons earned a score of 47. Similarly, the graph indicates that 20 persons earned a score of 46. Got it? Good!

Histogram

Histograms display data in the same way that line graphs do, except they use bars to represent each score instead of a line. To create a Histogram in SPSS, click on the *Graphs* menu and then on the *Histogram* item in the menu to

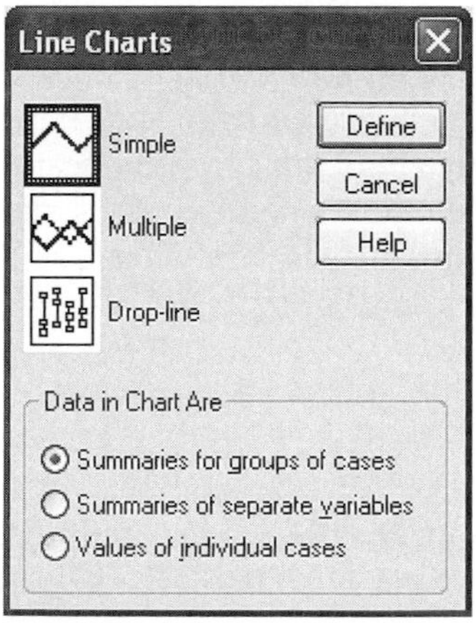

Fig 4-4 Line charts dialog box.

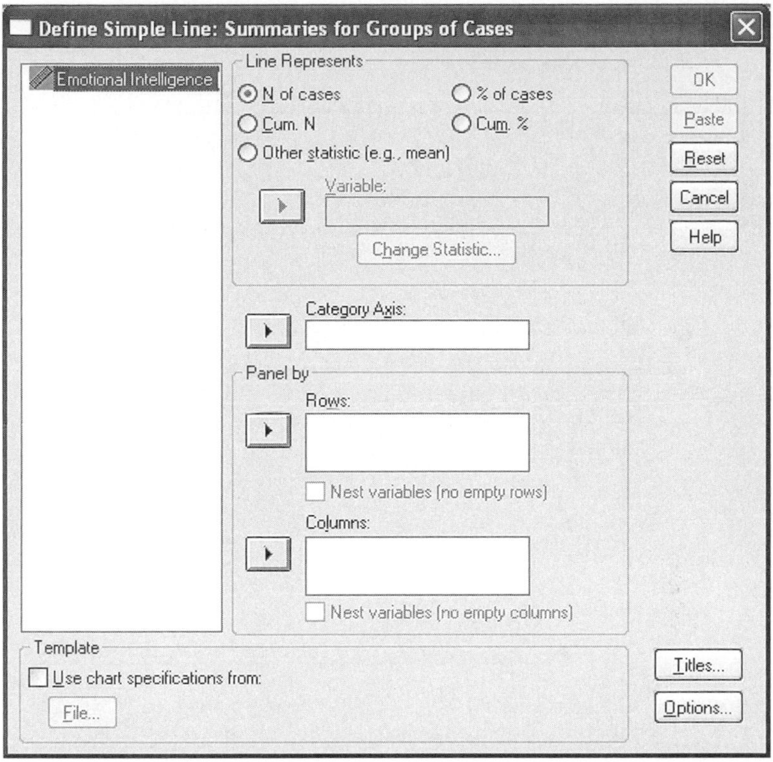

Fig 4-5 Define simple line: Summaries for groups of cases dialog box.

open the *Histogram* dialog box shown in Figure 4-7. Just as we did for creating a frequency polygon, click on the Emotional Intelligence variable in the dialog box and then on the ▶ next to the *Variable* area. Now click *OK* to create the histogram shown in Figure 4-8.

Cumulative Frequency Polygon

Occasionally, a researcher may want to display data in cumulative form. Instead of building a graph that shows the number of scores occurring at each possible score, a cumulative frequency polygon shows the number of scores occurring at or below each point. The steps for creating a cumulative frequency polygon in SPSS are the same as that for creating a line graph, except that *Cum. N* in the *Line Represents* area of the dialog box is selected. The cumulative frequency polygon for the emotional intelligence scores is shown in Figure 4-9.

By finding the point on the line that is exactly above any number on the horizontal axis, and then reading across to the left, we can see how many

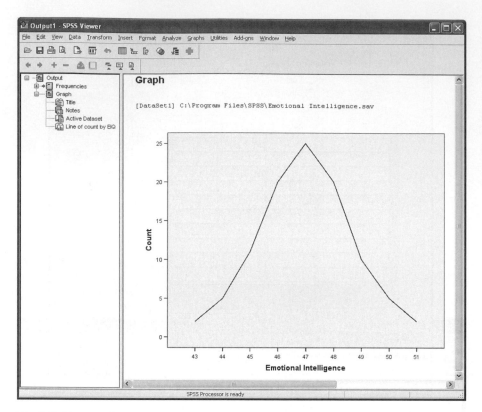

Fig 4-6 Line graph.

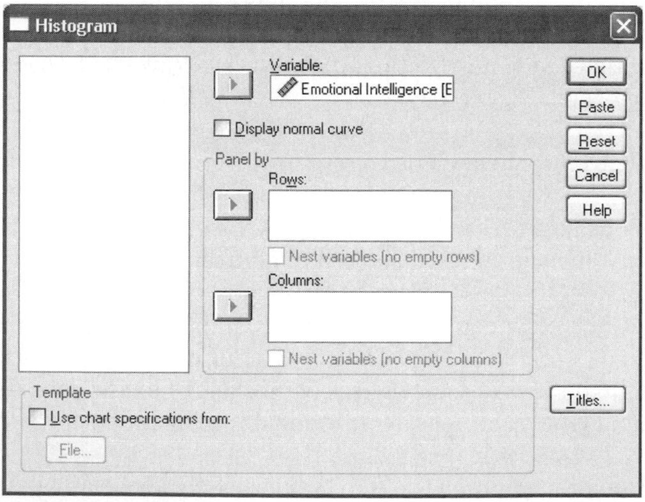

Fig 4-7 Histogram Dialog box

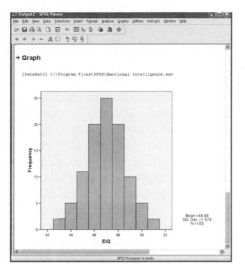

Fig 4-8 Histogram.

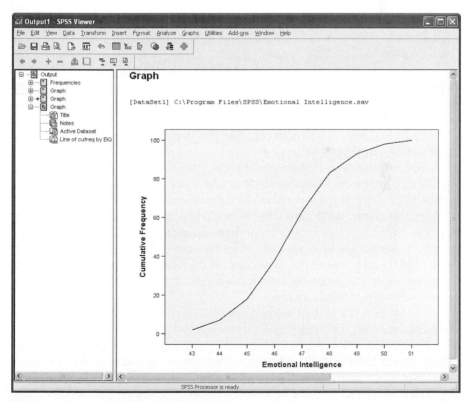

Fig 4-9 Cumulative frequency polygon.

students scored at or below that point. For example, 38 students had emotional intelligence scores at or below 46.

A cumulative frequency polygon is particularly useful in illustrating learning curves, when a researcher might be interested in knowing how many trials it took for a subject to reach a certain level of performance. Imagine that the graph here was obtained by counting the number of rounds a beginning dart thrower used during a series of practice sessions. The vertical axis is still "frequencies," but now it represents the number of practice rounds; the horizontal numbers represent the thrower's score on any given round. His worst score was 43, and he had two rounds with that score. He had 7 rounds with scores of either 43 or 44. How many rounds did he throw with scores of 47 or less? Well, find the point on the graph that is right over 47, and trace over to the left for the answer: 63 rounds yielded a score at or below 47. This cumulative graph is typical of a learning curve: The learning rate is relatively slow at first, picks up speed in the middle, and levels out at the end of the set of trials, producing a flattened S shape.

THE NORMAL CURVE

Around 1870, Adolphe Quetelet, a Belgian mathematician, and Sir Francis Galton, an English scientist, made a discovery about individual differences that impressed them greatly. Their method was to select a characteristic, such as weight or acuteness of vision, obtain measurements on large numbers of individuals, and then arrange the results in frequency distributions. They found the same pattern of results over and over again, for all sorts of different measurements. Figure 4-10 is an example that depicts the results of measuring chest size of over 5,000 soldiers.

The rectangles in this graph represent the number of folks who fell into each respective range. About 50 soldiers had chest sizes between 33.5 and 34.4 inches. If we were to put a mark at the top of each bar and draw straight lines between the marks, we'd have a frequency polygon of the sort we created earlier. The curve that's drawn over the bars doesn't follow that polygon shape exactly, however; it's what we'd get if we measured thousands and thousands more soldiers and plotted the histogram or frequency polygon for all of them, using narrower measurement intervals—maybe tenths or even hundredths of an inch—instead of whole inches.

The symmetrical, bell-shaped curve that results from plotting human characteristics on frequency polygons closely resembles a curve, familiar to mathematicians, known as the normal probability curve. The normal curve is bell-shaped and perfectly symmetrical and has a certain degree of "peakedness." Not all frequency distributions have this shape, however. Because of the importance of the normal curve to statistics, we will discuss it in further detail in Chapter 6.

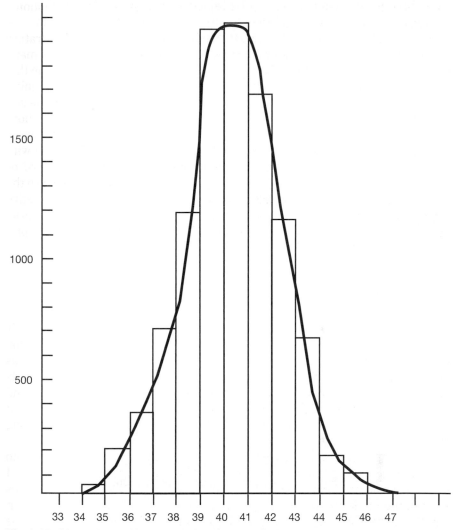

Fig 4-10 Chest sizes of 5,738 soldiers. Francis Galton, Natural Inheritance. London: Macmillan and Co., 1889.

Skewed Distributions

Skewed curves are not symmetrical. If a very easy arithmetic test were administered to a group of graduate students, for example, chances are that most students would earn high scores and only a few would earn low scores. The scores would tend to "bunch up" at the upper end of the graph, as if you'd taken a normal curve and pulled its bottom tail out to the left. When scores cluster near the upper end of a frequency polygon, so that the left side is "pulled down," the graph is said to be "negatively skewed." An example of a negatively skewed curve is shown in Figure 4-11. On the other hand, if the test were too difficult for the class, most people would get low scores. When graphed as a frequency polygon, these scores would be said to be "positively skewed," as shown in Figure 4-12.

PROBLEMS

1. Imagine that you are interested in studying the effects of type of preparatory instructions on reading fluency. In this experiment, the instructions would be the _____ variable and the number of words read correctly in one minute would be the _____ variable.
   (a) independent; dependent
   (b) dependent; dependent
   (c) dependent; independent
   (d) independent; independent
2. On a test of verbal ability, Mary obtained a score of 30, Bill a score of 45, and Sam a score of 60. If the difference between Mary's and Bill's scores is equivalent to

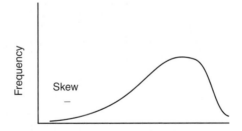

Fig 4-11 A negatively skewed distribution.

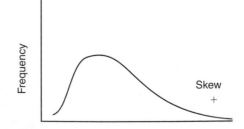

Fig 4-12 A positively skewed distribution.

the difference between Bill's and Sam's scores, then the level of measurement for these scores must be at least
(a) nominal
(b) ordinal
(c) interval
(d) ratio

3. Private, Corporal, Sargeant, Lieutenant form what kind of scale?

4. Suppose an exam was very easy and all but a few students obtained a high grade on it. The frequency distribution of these scores would be _____.
(a) negatively skewed
(b) positively skewed
(c) symmetrical
(d) bimodal

5. The normal distribution
(a) is bell-shaped.
(b) is important for the use of statistics.
(c) is symmetrical and asymptotic.
(d) all of the above.

6. The following scores were obtained by third graders on a weekly spelling test (10 points possible):

Spelling Test Scores									
4	3	10	3	3	2	9	3	8	3
2	3	1	5	4	0	1	4	0	3
2	4	3	1	4	2	8	2	1	2
1	5	2	9	3	6	4	4	3	2
1	4	1	3	3	2	2	2	8	3
9	4	9	3	3	10	1	3	5	3
2	2	4	3	3	6	6	4	1	2
6	2	3	7	4	4	4	4	2	4

(a) Construct a frequency distribution for these test scores.
(b) Graph the data using a frequency polygon. How would you characterize the resulting distribution of scores? Normal? Positively skewed? Negatively skewed?

ANSWERS TO PROBLEMS

1. a
2. c
3. Ordinal scale
4. a
5. d

6. a.

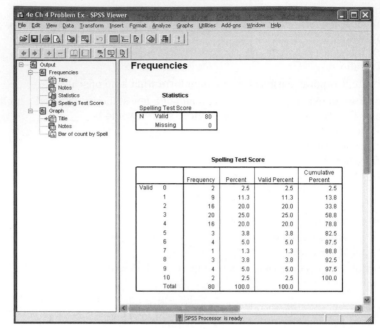

b.

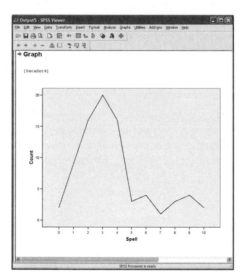

Positively Skewed.

5

Descriptive Statistics

- What are Statistics?
- Measures of Central Tendency
- Selecting a Measure of Central Tendency
- Measures of Variability
- Calculating Descriptive Statistics in SPSS
- Problem

There was this statistics student who, when driving his car, would always accelerate hard before coming to any intersection, whizz straight through it, and then slow down again. One day, he took a passenger, who was understandably unnerved by his driving style and asked him why he went so fast over junctions. The statistics student replied, "Well, statistically speaking, you are far more likely to have an accident at an intersection, so I just make sure that I spend less time there."

WHAT ARE STATISTICS?

Frequency distributions are very useful for providing a pictorial view of a set of data. While helpful, they often do not provide us with enough or with the right kind of information. We often ask questions such as, "What is the average GRE score of this class?" or "How much money does the average football player make?" When we ask such questions, we are really asking for a single number that will represent all of the different GRE scores, or player salaries, or whatever, rather than for the shape of a distribution. In such instances, measures of central tendency and variability—descriptive statistics—can be calculated to quantify certain important characteristics of a distribution. Many people are not aware that there is more than one "average." In this chapter, we will discuss three methods for computing an average: the mean, the median, and the mode. Just one number, though, can be misleading. Two very different sets of data can have the same average, yet differ greatly in terms of how much the scores vary by dataset. The second kind of summarizing technique is finding a number that describes that variability. Variability in a distribution can be described in terms of the range, variance (SD^2), or standard deviation (SD).

MEASURES OF CENTRAL TENDENCY

Measures of central tendency provide an average or typical score that describes the level (from high to low) of a set of scores. Measures of central tendency are useful when comparing the level of performance for a group of individuals with that of another group (e.g., boys vs. girls) or with some standard (e.g., national average), or when comparing the performance of the same group over time (e.g., before and after an intervention or treatment). The three main measures of central tendency are the mean, median, and mode.

Mean

The *mean* is the most often used measure of *central tendency* (a fancy statistical term that means, roughly, "middleness"). The mean is an old acquaintance of yours: the arithmetic average. You obtain the mean by adding up all the scores and dividing by the number of scores. Remember? Different statistics texts use different symbols to designate the mean. The most widely used method is to use a bar over the letter symbolizing the variable. For example, a group's mean score on variable X would be symbolized $\overline{X}$; the mean on variable Y would be $\overline{Y}$, and so on. By convention, the $\overline{X}$ and $\overline{Y}$ are used to designate the mean of a sample, that is, a finite set of something—test scores, heights, reaction times, what have you.

Sometimes we want to refer to the mean of a less definite, often infinite set: all the fifth graders in the United States, for example, or the scores that all those fifth graders would get if they all were given the same achievement test. A large, inclusive group like this is called a population; and the Greek letter μ (pronounced "mew," like the sound a kitten makes) symbolizes its mean. Values having to do with populations are called *parameters* and are usually symbolized using lowercase Greek letters; for sample values (called statistics), we use the normal English-language alphabet. To be technically correct, we would have to define a population as the collection of all the things that fit the population definition and a sample as some specified number of things selected from that population. You'll see why that's important when we talk about *inferential statistics* in Chapter 10. For now, though, just assume that we are working with samples—relatively small groups of things in which each individual member can be measured or categorized in some way.

The formula for the mean for variable X, or $\overline{X}$, is

$$\overline{X} = \frac{\Sigma X}{N}$$

where Σ means "the sum of," *X* refers to each obtained score, and *N* refers to the total number of scores.

Have you noticed how complicated it was to describe the mean in words, compared with that short little formula? Formulas, and mathematical relationships in general, often don't easily translate into words. Mathematicians are trained to think in terms of relationships and formulas and often don't have to translate; we do. That's one reason why social science folks can have problems with statistics: We don't realize that we need to translate and that the translating takes time. We expect to read and understand a page in a statistics book as quickly as a page in any other sort of book. Not so! Symbols simply take longer to read, and we need to remember to slow ourselves down. So slow down and take as long as you need to feel comfortable with a formula.

Median

When scores are arranged in order, from highest to lowest (or lowest to highest), the *median* (Mdn) is the middle score. In other words, the median is the score that divides the frequency distribution in half. Fifty percent of the total number of obtained scores fall above the median and 50% below. Suppose you administered a test to five persons who scored as follows:

$$113, 133, 95, 112, 94$$

To find the median, you would first arrange all scores in numerical order and then find the score that falls in the middle. Arranged from highest to lowest, these scores are

$$133, 113, 112, 95, 94$$

Here, the Mdn = 112, because two scores are higher than 112 and two scores are lower than 112. Finding the Mdn is easy when you have an odd number of scores. But what do you do when you have an even number? Suppose you have the following six scores:

$$105, 102, 101, 92, 91, 80$$

In this example, the number 101 can't be the Mdn, because there are two scores above it and three below. Nor can the number 92 be the Mdn, because there are three scores above and two below. With an even number of scores, the Mdn is defined as the point half the distance between the two scores in the middle. In this example, the two middle scores are 101 and 92. You find the point halfway between by adding the two middle scores and dividing by 2: 101 + 92 = 193, divided by 2 = 96.5 = Mdn. (Did you notice that this is the same as finding the mean of the two middle scores? Good for you!) The Mdn of our six scores is 96.5. As you can see, now there are three scores that are higher than 96.5 and three that are lower.

Mode

The *mode* (Mo) is simply the most frequently occurring score in a set of scores. For example, suppose we are given the following scores:

$$110, 105, 100, 100, 100, 100, 99, 98$$

Because the number 100 occurs more frequently than any of the other scores, Mo = 100. Simple enough? But what about the following set of scores?

$$110, 105, 105, 105, 100, 95, 95, 95, 90$$

In this example, both 105 and 95 occur three times. Here, we have a distribution with two modes: a bimodal distribution. If there were more than two modes, it would be called a multimodal distribution.

SELECTING A MEASURE OF CENTRAL TENDENCY

Mark Twain once said, "There are three kinds of lies: Lies, damned lies, and statistics." Actually, statistics don't lie. But they can be employed to enhance communication—or to deceive those who do not understand their properties. (Although we wouldn't want to do that, would we?) For example, consider the University of Football, a university so small that its entire staff consists of five persons, four professors and a football coach. Their annual incomes are as follows:

Professors' Salary	Coach's Salary
Professor 1 $60,000	$1,000,000
Professor 2 $55,000	
Professor 3 $55,000	
Professor 4 $50,000	

The coach boasts that the university is a fantastic university with an "average" annual salary of $244,000. Before rushing off to join the faculty, you would be wise to find out what measure of central tendency he is using! True, the mean is $244,000, so the coach is not lying. But is that the typical salary of all employees at the University of Football? Of course not. The coach's income is quite extreme in comparison to that of his fellow employees, so the story he is telling is not very accurate. In this case, either the median or the mode would be a more representative value of the typical salary of employees. For both, the "average" income is $55,000. The point made here is that your selection of a measure of central tendency will be determined by your objectives in communication as well as by mathematical considerations.

The mean is the only measure of central tendency that reflects the position of each score in a distribution. Because the mean is the "balance point" of the distribution, the mean is affected most by extreme scores. The mean is often the best choice for the average when the distribution of scores is symmetrical. The Mdn, in contrast, responds to how many scores lie above and below it, not how far above and below. It doesn't matter whether the coach's income in the example above is $61,000 or $1,000,000 when determining the Mdn. Because the Mdn is less affected by extreme scores, the Mdn is often the best choice of the average when the distribution is skewed. The mode, although easy to obtain, is really only suitable when you want to know the "most likely" value in a distribution. The mode is also the only measure of central tendency that can be used with scores obtained on a nominal scale of measurement.

When a distribution is skewed, the mean is the measure of central tendency that is most strongly affected. A few scores far out in the tail of a distribution will "pull" the mean in that direction. The median is somewhat "pulled" in the direction of the tail, and the mode is not "pulled" at all (see Figure 5-1). To see what I mean, look at the three distributions shown in Figure 5-2. The first distribution (X) is perfectly symmetrical. Its mode, mean, and median are equal. In unimodal, symmetrical distributions, $\overline{X} = \text{Mdn} = \text{Mo}$, always. Distribution Y has been skewed by changing the largest score, 5, to 12. Skewing it to the right like this shifted the mean from 3 to 4, but it didn't

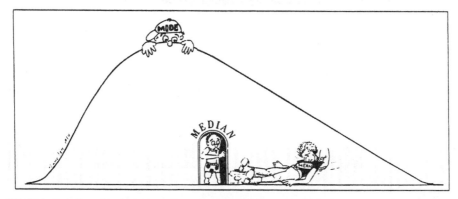

Fig 5-1 Relationship between the mean, median, and mode in a positively skewed distribution.

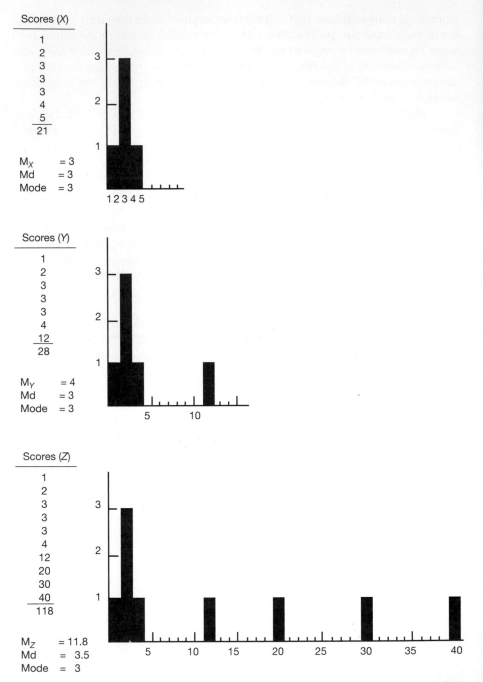

Fig 5-2 Symmetrical and skewed distributions.

change either the mode or the median. Finally, in distribution *Z*, not only have we shifted the 5 out to 12 but we have further skewed the distribution to the right by adding scores of 20, 30, and 40. The mean is again most sensitive to these changes. And this time, the Mdn shifts too, just a bit, from 3 to 3.5. But the mode remains unchanged.

MEASURES OF VARIABILITY

When we computed measures of central tendency (mean, median, and mode), we were looking for one score that would best represent the level of an entire set of scores. Consider the final exam scores earned by students in each of two statistics classes:

Classroom X Exam Scores	Classroom Y Exam Scores
160	102
130	101
100	100
70	99
40	98
$\Sigma X = 500$	$\Sigma Y = 500$
$\overline{X} = \dfrac{500}{5} = 100$	$\overline{Y} = \dfrac{500}{5} = 100$

Notice that the mean of both classrooms is 100. But what a difference in variability! (Perhaps you have heard about the man who drowned in a lake with an average depth of 1 foot.) In order to deal with such differences, statisticians have developed several measures of variability that allow us to differentiate between groups of scores like these. Whereas measures of central tendency describe the level of a set of scores, measures of variability describe the differences among a set of scores. In other words, they provide an estimate of how much scores in a set are spread out or clustered together. The measures of variability we will discuss are the range, variance (SD^2), and standard deviation (SD).

Range

The simplest measure of variability is the range. The range is the highest score (H) minus the lowest score (L).

$$\text{Range} = \text{H} - \text{L}$$

In Classroom X,

$$\text{Range} = \text{H} - \text{L} = 160 - 40 = 120$$

In Classroom Y,

$$\text{Range} = \text{H} - \text{L} = 102 - 98 = 4$$

Because the range is based on the two most extreme scores, it can be quite misleading as a measure of overall variability. Remember the University of Football, where the coach had an annual income of $1,000,000 and the other four professors had incomes of $60,000, $55,000, $55,000, and $50,000? The range of this distribution is $950,000, even though all but one of the people in the sample are clustered within $10,000 of each other. In this distribution, the range is not as useful a measure as the variance and standard deviation, which are based on all the scores. That's where we go next. First, let's talk about the variance.

Variance (SD^2)

The variance (SD^2) is the most frequently used measure of variability and is defined as the mean of the squares of deviation scores. The formula for the variance may seem a little bit intimidating at first, but you can handle it if you follow the procedures outlined in what follows:

$$SD^2 = \frac{\Sigma(X - \bar{X})^2}{N - 1}$$

Where: Σ means "the sum of,"
           X refers to each obtained score,
           $\bar{X}$ is the mean of X, and
           N refers to the total number of scores.

Before learning how to compute the variance with SPSS, let's discuss the concept of deviations from the mean, or deviation scores.

Deviation Scores

We can tell how far each score deviates from the mean by subtracting the mean from it, using the formula $x = (X - \bar{X})$. Positive deviation scores indicate positions that are above the mean and negative deviation scores indicate positions below the mean. Notice that for the following scores, we have sub-tracted the mean of 5.0 from each score:

Scores (X)	($X - \bar{X}$)	χ
9	9 − 5	+4
7	7 − 5	+2
5	5 − 5	0
3	3 − 5	−2
1	1 − 5	−4
$\Sigma X = 25$		$\Sigma(X - \bar{X}) = \chi = 0$

$$\bar{X} = \frac{\Sigma X}{N} = \frac{25}{5} = 5.0$$

All we have to do to find how much these scores differ from the mean on aver-age is to calculate the mean deviation score, right? Unfortunately, it's not that easy. As you can see, when we add up the column headed "$X - \bar{X}$" the sum of that column equals zero. Because the mean is the "balance point" of a set of scores, the sum of deviations about their mean is always zero, except when you make rounding errors. In fact, another definition of the mean is the score around which the sum of the deviations equals zero. The mean deviation score is not a very good measure of variability, therefore, because it is the same for every distribution, even when there is wide variability among distributions. In our example, we can easily see that they do in fact differ.

So how do we estimate variability? This is where the variance comes to the rescue. If we square each deviation score, the minus signs cancel each other out. In the following distribution, look carefully at the column headed $(X - \bar{X})^2$. Notice that by squaring the deviation scores we get rid of the nega-tive values.

Scores (X)	($X - \bar{X}$)	χ	($X - \bar{X}$)^2
9	9 − 5	+4	16
7	7 − 5	+2	4
5	5 − 5	0	0
3	3 − 5	−2	4
1	1 − 5	−4	16
$\Sigma X = 25$	$\Sigma(X - \bar{X}) = \chi = 0$		$\Sigma(X - \bar{X})^2 = 40$

Now, for the sum of the squared deviations, we have $\Sigma(X - \bar{X})^2 = 40$, the numerator of the formula for the variance. To complete the computation for the variance, just divide by $N - 1$:

$$SD_x^2 = \frac{\Sigma(X - \bar{X})^2}{N - 1} = \frac{40}{5 - 1} = 10$$

The variance is defined as the mean of the *squared deviation* scores. In other words, the variance is a kind of average of how much scores deviate from the mean after they are squared. The difference between this formula and other formulas for "average" values is, of course, that here we divide by $N - 1$ rather than simply by the number of scores. The reason? It's fairly complicated, and it has to do with how we will use the variance later on, when we get to *inferential statistics*. I think it would be confusing to explain it now, so let's wait to talk about this until we get to Chapter 10.

Many students complete the computation of their first variance and then ask, "What does it mean?" "What good are squared deviation scores?" Perhaps

THE FAR SIDE By GARY LARSON

"Yes, yes, I *know* that, Sidney—*every*body knows *that*! ... But look: Four wrongs *squared*, minus two wrongs to the fourth power, divided by this formula, *do* make a right."

you have a similar question. To statisticians, the variance reflects the "amount of information" in a distribution.

In any case, let's go back to Classrooms X and Y (from the beginning of the chapter) and find the variance for each classroom. The variance for Classroom X is

$$SD_x^2 = \frac{\Sigma(X - \overline{X})^2}{N - 1} = \frac{9,000}{5 - 1} = 2,250$$

And the variance for Classroom Y is

$$SD_y^2 = \frac{\Sigma(Y - \overline{Y})^2}{N - 1} = \frac{10}{5 - 1} = 2.5$$

Notice that the values for the variances of Classrooms X and Y reflect the fact that there is quite a bit of difference between the variabilities of the classrooms. That's what the variance is supposed to do—provide a measure of the variability. The more variability in a group, the higher the value of the variance; the less variability in a group, the lower the value of the variance.

Another way to understand the variance is to notice that, in order to find it, you need to add all the squared deviations and divide by $N - 1$. Sound familiar? Very similar to the definition of the mean, don't you think? So one way to understand the variance is to think of it as an average deviation squared, or maybe a mean squared deviation.

Standard Deviation (*SD*)

When you read educational and psychological research, you will often come across the term *standard deviation*. Once you have found the variance of a sample, finding the standard deviation is easy: Just take the square root of the variance. If the variance is 25, the standard deviation will be 5; if the variance is 100, the standard deviation will be 10. To find the standard deviations of the exam scores from Classrooms X and Y, take the square root of the variances.

SD of Classroom X	SD of Classroom Y
$SD_x = \sqrt{SD_x^2} = \sqrt{2,250} = 47.4$	$SD_Y = \sqrt{SD_Y^2} = \sqrt{2.5} = 1.6$

But what do the numbers mean? Again, as was true with the variance, the values of computed standard deviations indicate the relative variability within a group. From our descriptive statistics we know that the variability among students in

Classroom X is much greater than that among students in Classroom Y. What does this mean? Why are there such differences? These are interesting questions that statistics cannot answer for you. Such differences in variability could be due to differences among students across classes in prior knowledge of math and statistics; or they could be due to differences in teaching practices; or they could be due to something else.

CALCULATING DESCRIPTIVE STATISTICS IN SPSS

Calculating descriptive statistics in SPSS can be done in a couple of different ways. For these examples, we'll use the dataset in Figure 5-3.

As you can see in the figure, we have data for five students on an aptitude test, midterm examination, and their grade-point average (GPA). To calculate descriptive statistics for these variables, click on *Analyze* in the menu bar, then on the *Descriptive Statistics* item in the menu, and finally on *Frequencies* . . . to open the *Frequencies* dialog box shown in Figure 5-4.

As we did in Chapter 4, move each of the three variables to the *Variable(s):* box by clicking on each and then on the ▶ in the middle of the dialog box. Now click on the *Statistics* button to bring up the *Frequencies: Statistics* dialog box shown in Figure 5-5.

In this box select the measures of central tendency and dispersion that we have discussed, as shown in the figure, and then click on *Continue*. This takes you back to the Frequencies dialog box. If you want to see the frequency distribution

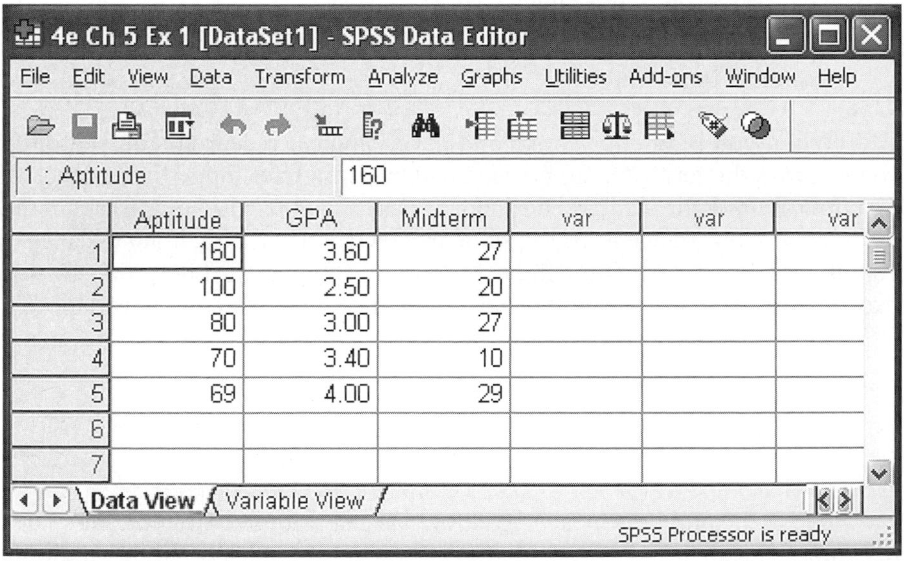

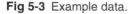

⛶ 4e Ch 5 Ex 1 [DataSet1] - SPSS Data Editor					_ □ ✕
File Edit View Data Transform Analyze Graphs Utilities Add-ons Window Help					

1 : Aptitude		160				
	Aptitude	GPA	Midterm	var	var	var
1	160	3.60	27			
2	100	2.50	20			
3	80	3.00	27			
4	70	3.40	10			
5	69	4.00	29			
6						
7						

◀ ▶ \ **Data View** ⋀ Variable View /

SPSS Processor is ready

Fig 5-3 Example data.

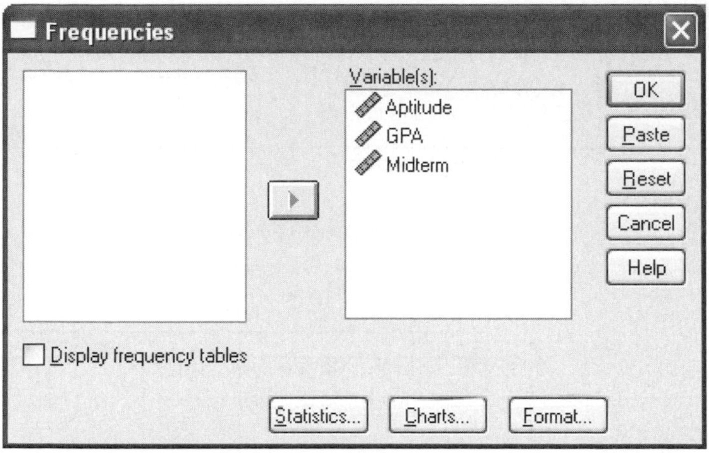

Fig 5-4 Frequencies dialog box.

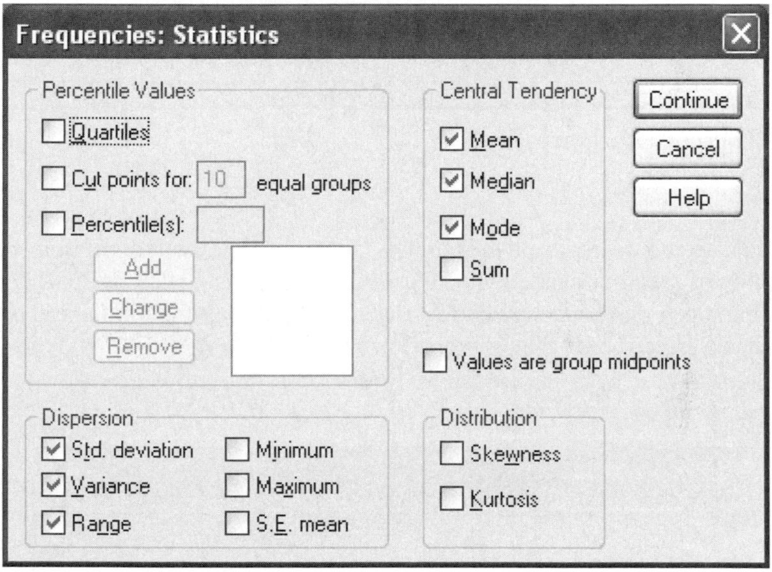

Fig 5-5 Frequencies: Statistics dialog box.

for these data, select *Display frequency tables*. If you do not, do not select this option. For this example, I have deselected the *Display frequency tables* option, because we have already covered how to create frequency tables. Once you have made a selection, click *OK* to run the analysis. Figure 5-6 displays the results.

As you can see here, the output consists of the descriptive statistics for our data. Please note that SPSS shows you the smallest value when you have multiple modes in your dataset.

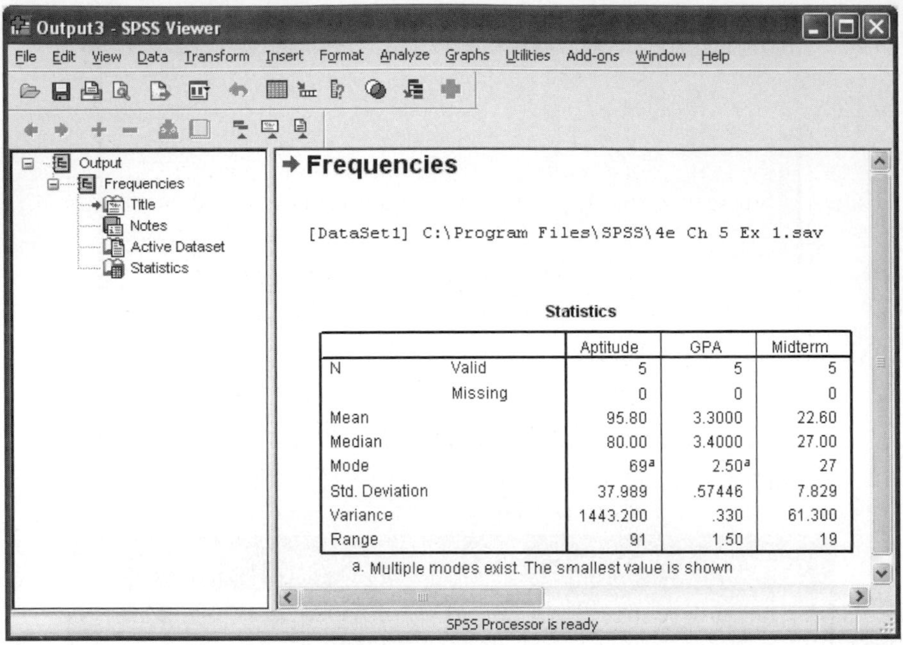

Fig 5-6 Descriptive statistics results.

You can also create descriptive statistics through their own menu item without the option of creating a frequency distribution. To do so, click on *Analyze* in the menu bar, then on the *Descriptive Statistics* item, and finally on *Descriptives . . .* to open the *Descriptives* dialog box shown in Figure 5-7.

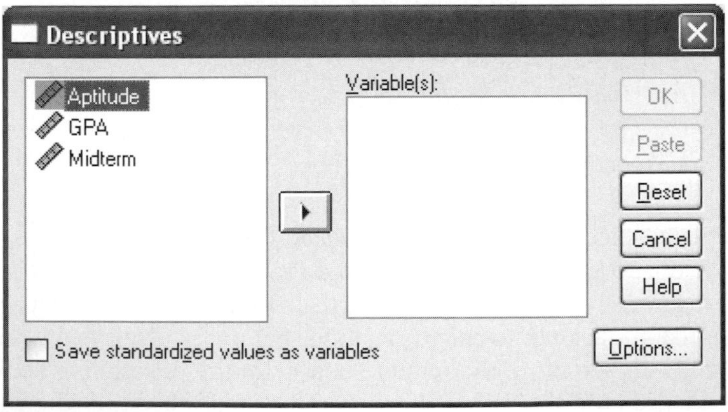

Fig 5-7 Descriptives dialog box.

In the dialog box, click on each of the variables in the list and then on the ▶ in the middle of the dialog box. This moves the variables we want to analyze to the *Variable(s)* box. Now click *Options . . .* to open the *Descriptives: Options* dialog box shown in Figure 5-8.

This box tells SPSS which descriptive statistics to calculate. Select the options for the mean, standard deviation, variance, and range, and click *Continue*. Please note that finding the mode is not an option when you calculate

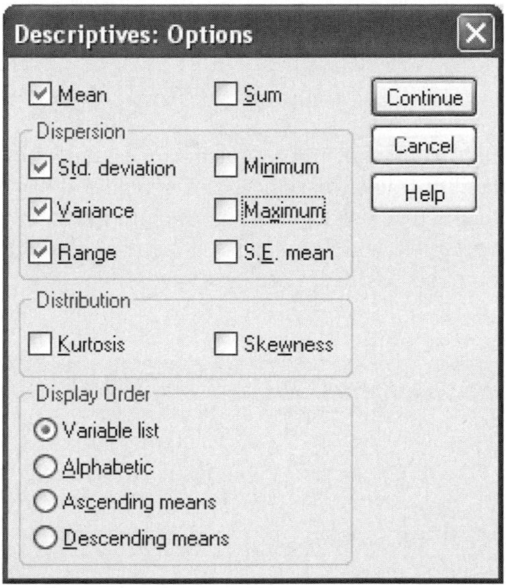

Fig 5-8 Descriptives: Options dialog box.

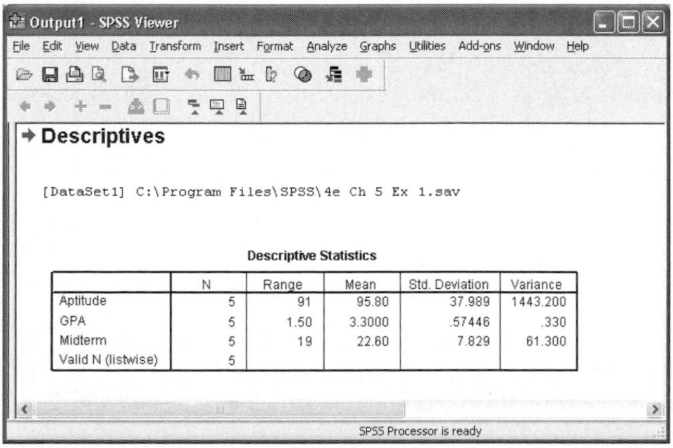

	N	Range	Mean	Std. Deviation	Variance
Aptitude	5	91	95.80	37.989	1443.200
GPA	5	1.50	3.3000	.57446	.330
Midterm	5	19	22.60	7.829	61.300
Valid N (listwise)	5				

Fig 5-9 Descriptive statistics results.

descriptive statistics this way. Once you are back in the *Descriptives* dialog box, click *OK* to run the analysis. Results of this analysis are shown in Figure 5-9.

As you can see, results of both analyses are identical. The only difference is the mode of presentation and the options that are available in the dialog box. Which one should you use? That's entirely up to you, unless you want to find the mode, in which case you will need to calculate descriptive statistics via the *Frequencies* menu item.

PROBLEM

1. A researcher wants to calculate measures of central tendency (mean, median, and mode) and variability (range, variance, and standard deviation) for measures of self-esteem and assertiveness for a group of eight gifted middle school children. For these two variables, which measure of central tendency should she use to describe the level of scores? Here are her data:

ANSWER TO PROBLEM

1. Because the distribution of scores for self-esteem is skewed, the researcher should use the median (Mdn = 64.0) as the best measure of central tendency. As you can see here, the mean and median differ substantially. For assertiveness, the mean (Mean = 4.6) can be used because there are no extreme scores. In addition, for both variables, the mode is not the most appropriate measure of

central tendency, because we are interested in knowing the level of scores, not the "most likely" score.

6

....

The Normal Curve

- What is the Normal Curve?
- Proportions of Scores under the Normal Curve
- Problems

A famous statistician would never travel by airplane because he had studied air travel and estimated the probability of there being a bomb on any given flight was 1 in a million, and he was not prepared to accept these odds. One day a colleague met him at a conference far from home. "How did you get here, by train?" "No, I flew." "What about the possibility of a bomb?" "Well, I began thinking that if the odds of one bomb are 1 in a million, then the odds of TWO bombs are $(1/1,000,000) \times (1/1,000,000)$. This is a very, very small probability, which I can accept. So, now I bring my own bomb along!"

The normal curve is a mathematical idealization of a particular type of symmetric distribution. It is not a fact of nature. Since many actual distributions in the social and behavioral sciences approximate the normal curve, we can use what mathematicians know about it to help us interpret test results and other data. When a set of scores is distributed approximately like the normal curve, mathematicians can provide us with a great deal of information about those scores, especially about the proportions of scores that fall in different areas of the curve. Before discussing how to use the normal curve to find certain proportions of scores, let's talk a little bit more about the characteristics of the normal curve.

WHAT IS THE NORMAL CURVE?

The normal distribution describes a family of normal curves, just like a circle describes a family of circles—some are big, some are small, but they all have certain characteristics in common. What are the common features of normal curves?

1. All normal curves are symmetric around the mean of the distribution. In other words, the left half of the normal curve is a mirror image of the right half.

2. All normal curves are unimodal. Because normal curves are symmetric, the most frequently observed score in a normal distribution—the mode—is the same as the mean.

3. Since the normal curves are unimodal and symmetric, the mean, median, and mode of all normal distributions are equal.

4. All normal curves are asymptotic to the horizontal axis of the distribution. Scores in a normal distribution descend rapidly as one moves along the horizontal axis from the center of the distribution toward the extreme ends of the distribution, but they never actually touch it. This is because scores on the normal curve are continuous and held to describe an infinity of observations.

5. All normal curves have the same proportions of scores under the curve relative to particular locations on the horizontal axis when the scores are expressed in a similar basis (i.e., in standard scores).

SidneyHarris/www.sciencecartoonsplus.com

All normal curves have these features in common, but they can differ in terms of their mean and standard deviation.

The Normal Curve as a Model

As mentioned above, the normal curve is a good description of the distribution of many variables in the social and behavioral sciences. The distribution of scores on standardized tests of intelligence, or "IQ" tests, for example, is roughly normal. Not all variables are normally distributed, however. The distribution of reaction time, for example, is positively skewed. This is because there is a physiological lower limit to the time in which a person can respond to a stimulus, but no upper limit.

In addition to the fact that the normal curve provides a good description for many variables, it also functions as a model for distributions of statistics. Imagine that you randomly selected a sample of people from a population, calculated the mean of that sample, and then put them back into the population. Now imagine doing that again and again, an infinite number of times. If you created a frequency distribution of the means for those samples, the distribution of those statistics would approximate the normal curve. This is known as a sampling distribution. Knowing the shape of distributions of statistics is crucial for inferential statistics, as we'll see in later chapters in Section IV.

PROPORTIONS OF SCORES UNDER THE NORMAL CURVE

In the next few paragraphs, I am going to show you how the mean, median, mode, standard deviation, and the normal probability curve are all related to each other. Consider the following example:

Test Scores (X)	Frequency
110	1
105	2
100	3
95	2
90	1

If you were to calculate the mean, median, and mode from the data in this example, you would find that $\overline{X} = \text{Mdn} = \text{Mo} = 100$. Go ahead and do it, just for practice. The three measures of central tendency always coincide in any group of scores that is symmetrical and unimodal.

Recall that the median is the middle score, the score that divides a group of scores exactly in half. For any distribution, you know that 50% of the remaining scores are below the median and 50% above. In a normal distribution, the median equals the mean, so you know that 50% of the scores are higher than the mean and 50% are lower. Thus, if you know that the mean of a group of test scores is 70, and if you know that the distribution is normal, then you know that 50% of the persons who took the test (and who didn't get a score of exactly 70) scored higher than 70 and 50% lower.

Now let's see how the standard deviation fits in. Suppose again that you had administered a test to a very large sample, that the scores earned by that sample were distributed like the normal probability curve, and that the $\overline{X}$ = 70 and the SD_x = 15. Mathematicians can show that in a normal distribution, exactly 68.26% of the scores lie between the mean and 1 standard deviation away from the mean. (You don't need to know why it works out that way; just take it on faith.) In our example, therefore, about 34.13% of the scores would be between 70 and 85 (85 is 1 standard deviation above the mean: $\overline{X} + SD_x$ = 70 + 15 = 85). We know that the normal curve is symmetrical, so we know that about 34.13% of the scores will also be between 70 and 55 ($\overline{X} - SD_x$ = 70 − 15 = 55). Thus, if we administered our test to 100 persons, approximately 34 would have scores between 70 and 85, and about 68 would have scores between 55 and 85 (34.13% + 34.13%). Most students find it helpful (necessary?) to see a picture of how all of this works; use the graph in Figure 6-1 to check it out.

To take another example, suppose that grade-point averages (GPAs) were calculated for 1,000 students and that for GPA the $\overline{X}$ = 2.5 and SD_x = 0.60. If our sample of GPAs was drawn from a normal distribution, you would know that approximately 500 students had GPAs higher than 2.5 and 500 had GPAs lower than 2.5. You would also know that approximately 683 of them (34.13% + 34.13% = 68.26%; 68.26% of a thousand is approximately 683) had GPAs somewhere between 1.90 and 3.10 (the mean ±1 standard deviation). Figure 6-2 shows what the graph would look like.

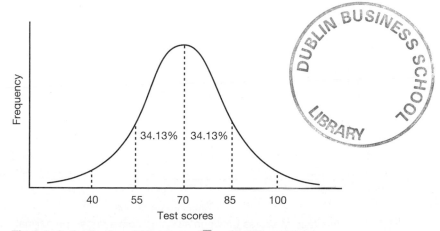

Fig 6-1 A distribution of IQ scores ($\overline{X}$ = 70, SD_x = 15).

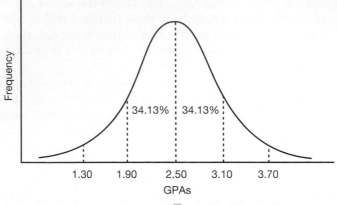

Fig 6-2 A distribution of GPAs ($\overline{X} = 2.50$, $SD_x = 0.60$).

Mathematicians can tell us the proportion of the population between any two points under the normal curve, because the area under the normal curve is proportional to the frequency of scores in the distribution. Figure 6-3 presents information about some selected points. The numbers on the baseline of the figure represent standard deviations (SD), where -1 represents 1 standard deviation below the mean, $+2$ represents 2 standard deviations above the mean, and so on.

Given the information in the graph, you can answer some interesting questions. For example, suppose again that you administered a test to 100 people, that the scores were distributed approximately normally, and that $\overline{X} = 70$ and $SD_x = 15$. You know from the preceding discussion that about half

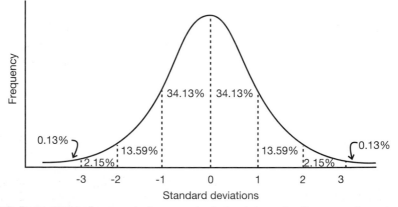

Fig 6-3 Percentage of scores between selected points under the normal curve.

of the 100 people scored lower than 70 and half of them scored higher; and you know that approximately 68 scored between 55 and 85. The graph also tells you the following:

1. About 13.59% of the scores are between $-1SD$ and $-2SD$. In our example, therefore, around 14 people scored between 40 (2 standard deviations below the mean) and 55 (1 standard deviation below the mean). Of course, 13.59% (approximately) of the cases also are between $+1SD$ (85) and $+2SD$ (100).
2. About 95.44% of the cases fall between $-2SD$ and $+2SD$ (13.59% + 34.14% + 34.13% + 13.59%), so we know that approximately 95 out of 100 people scored between 40 ($-2SD$) and 100 ($+2SD$).
3. About 99.74% of the cases fall between $-3SD$ and $+3SD$. Virtually all persons in our example scored between 25 ($-3SD$) and 115 ($+3SD$).
4. About 84.13% had scores lower than 85. How do we know this? Fifty percent of the persons scored below 70 (the mean and median, remember?), and another 34.13% scored between 70 and 85. Adding the 50% and the 34.13%, you find that 84.13% of the scores are predicted to fall below 85.

The following problems will test your understanding of the relationships among the mean, median, mode, standard deviation, and the normal probability curve. Remember that these relationships hold only if the data you are working with are distributed normally. When you have a skewed distribution or one that is more or less peaked than normal, what you have just learned does not hold true.

PROBLEMS

1. Suppose that a test of math anxiety was given to a large group of persons, the scores are assumed to be from a normally distributed population, and that $\bar{X} = 50$ and $SD_x = 10$. Approximately what percentage of persons earned scores:
   (a) below 50?
   (b) above 60?
   (c) between 40 and 60?
2. A major pharmaceutical company has published data on effective dosages for their new product, FeelWell. It recommends that patients be given the minimum effective dose of FeelWell and reports that the mean effective minimum dose is 250 mg, with a standard deviation of 75 mg (dosage effectiveness is reported to be normally distributed). What dose level will be effective for all but about 2% of the total population? What dose level can be expected to be too low for all but approximately 2%?

3. New American cars cost an average of $17,500, with $SD = \$2,000$. (That's not really true; I just made it up for this problem.) If I'm only willing to spend up to $15,500, and if car prices are normally distributed, what percentage of the total number of new cars will fall within my budget?

ANSWERS TO PROBLEMS

1. **(a)**   50%   **(b)**   15.87% **(c)**   68.26%
2. 400 mg; 100 mg
3. 15.87%

7

Percentiles and Standard Scores

- Percentiles
- Standard Scores
- Other Standard Scores
- Converting Standard Scores to Percentiles
- A Word of Warning
- Problems

A musician drove his statistician friend to a symphony concert one evening in his brand new mid-sized Chevy. When they arrived at the hall, all the parking spots were taken except one in a remote, dark corner of the lot. The musician quickly maneuvered his mid-sized Chevy into the space and they jumped out and walked toward the hall. They had only taken about ten steps when the musician suddenly realized he had lost his car key. The statistician was unconcerned because he knew the key had to be within 1 standard deviation of the car. They both retraced their steps and began searching the shadowed ground close to the driver's door. After groping on his hands and knees for about a minute, the musician bounced to his feet and bolted several hundred yards toward a large street light near the back of the concert hall. He quickly got down on all fours and resumed his search in the brightly lit area. The statistician remained by the car dumbfounded knowing that the musician had absolutely zero probability of finding the key under the street light. Finally, after 15 minutes, the statistician's keen sense of logic got the best of him. He walked across the lot to the musician and asked, "Why in the world are you looking for your key under the street light? You lost it back in the far corner of the lot by your car!" The musician in his rumpled and stained suit slowly got to his feet and muttered angrily, "I KNOW, BUT THE LIGHT IS MUCH BETTER OVER HERE!!"

If you work as a teacher or a member of one of the other helping professions (e.g., school psychologist, psychologist, counselor) you frequently will be required to interpret material in student or client folders. Material in the folders typically will include several types of test scores. This chapter will introduce you to two common types of scores—percentiles and standard scores—as well as their major variations. Some of the material may appear complicated at first, but it's just a logical extension of what you've learned so far. You may not even find it particularly difficult!

Before discussing percentiles and standard scores, I want to point out some of the disadvantages of three types of scores with which you may be familiar from your school days: the raw score, the percentage correct score, and rank in class. Consider the following dialog:

Boy: Mom! I got 98 on my math test today!

Mother: That's very good, son. You must be very happy!

Boy: Yes, but there were 200 points on the test.

Mother: Oh! I'm sorry. I guess you didn't do too well.

Boy: Yes, but I got the second-highest score in class.

Mother: Very good!

Boy: Yes, but there are only two of us in the class.

As you can see, the number of points the boy obtained on the math test—his raw score—didn't communicate much information. But neither did his percentage correct score, because it didn't tell us whether the test was extremely difficult or very easy. Nor was his rank in class very helpful unless we knew how large the class was, and even when we found that out, we didn't know a whole lot, because the class was small and our knowledge of the one person

SidneyHarris/www.sciencecartoonsplus.com

with whom he was being compared is nonexistent. When interpreting some-one's test score, we would like to know, at a minimum, something about the group of persons with whom he or she is being compared (the norm group) and how he or she did in comparison with that group.

The norm group of a test usually will include a large sample of people. For a standardized aptitude test, for example, test publishers often attempt to get a large, representative sample of people in general. They often include a sample of persons in the various age, gender, racial/ethnic, and socioeconomic groups likely to be measured by that test, in proportions reflecting the United States Census data. Test manuals often contain detailed descriptions of normative samples and the way in which they were obtained; a good textbook on tests and measurement can also give you that kind of information. Obviously, a test's use-fulness to you is strongly influenced by the group(s) on which it was normed. For instance, if you intend to use a test with preschoolers, but it was normed on adults, you would have no way of meaningfully interpreting your clients' scores.

PERCENTILES

Percentiles are one of the most frequent types of measures used to report the results of standardized tests, and for good reason: They are the easiest kind of score to understand. An individual whose score is at the 75th percentile of a group scored higher than about 75% of the persons in the norm group; some-one whose score is at the 50th percentile scored higher than about 50% of the persons in the norm group; someone whose score is at the 37th percentile scored higher than about 37% of the persons in the norm group; and so on.

The percentile rank of a score in a distribution is the percentage of the whole distribution falling below that score, plus half the percentage of the dis-tribution falling exactly on that score. Consider the following two distributions:

Distribution A	Distribution B
1	4
2	5
3	5
4	5
5	5
5	5
6	5
7	5
8	5
9	10

In both distributions, a score of 5 is at the 50th percentile; it has a percentile rank of 50. In distribution A, 40% of the scores are below 5, and 20% are at 5.

In distribution B, 10% of the scores are below 5, and 80% are at 5. Fortunately, most scores are normed on very large groups in which the scores form an approximately normal distribution, so you can take a person's percentile rank as a very close estimate of how many folks could be expected to score lower than that person. It is important to note that if the percentile rank is based on a small group, or on one that isn't normally distributed, you will need to be more cautious in interpreting it.

No matter how large or small the group, though, or what the shape of the distribution, computing a score's percentile rank is always the same: the percentage of scores below the one in question, plus half the percent right at that score.

STANDARD SCORES

On many published psychological tests, raw scores are converted to what are called *standard scores*. Standard scores are very useful, because they convert raw scores to scores that are meaningfully interpreted. This makes it possible to compare scores or measurements from very different kinds of distributions. The most basic standard score is known as the Z score. Z scores state the position of a score in relation to the mean in standard deviation units. Let's see how it works.

The Z score formula is as follows:

$$Z = \frac{X - \bar{X}}{SD_x}$$

Where: X is an individual's raw score,
$\bar{X}$ is the mean of the group with which the individual is being compared (usually a norm group of some kind), and
SD_x is the standard deviation of that group.

With Z scores you can compare all scores, from any distribution, on a single, comparable basis. The Z score has the following properties:

1. The mean of any set of Z scores is always equal to 0.
2. The standard deviation of any set of Z scores is always equal to 1.
3. The distribution of Z scores has the same shape as the distribution of raw scores from which they were derived.

Suppose you administered a test to a large number of persons and computed the mean and standard deviation of the raw scores with the following results:

$$\bar{X} = 42$$
$$SD_x = 3$$

Suppose also that four of the individuals tested had these scores:

Person	Score (X)
Jim	45
Sue	48
George	39
Jane	36

What would be the Z score equivalent of each of these raw scores? Let's find Jim's Z score first:

$$Z_{\text{Jim}} = \frac{\text{Jim's Score} - \bar{X}}{SD_x} = \frac{45 - 42}{3} = \frac{3}{3} = +1$$

Notice that (a) we substituted Jim's raw score ($X = 45$) into the formula, and (b) we used the group mean ($\bar{X} = 42$) and the group standard deviation ($SD_x = 3$) to find Jim's Z score. Because lots of Z scores turn out to have negative values, we use the $+$ sign to call attention to the fact that this one is positive.

Now for George's Z score:

$$Z_{\text{George}} = \frac{\text{George's Score} - \bar{X}}{SD_x} = \frac{39 - 42}{3} = \frac{-3}{3} = -1$$

Your turn—you figure out Sue's and Jane's Z scores. Did you get $Z_{\text{Sue}} = +2$, and $Z_{\text{Jane}} = -2$? If you did, you've already got the hang of it. Nice job!

Fortunately, SPSS makes the task of calculating Z scores very simple. For example, assume we have the scores of 15 participants in a study on long-term memory recall shown in Figure 7-1.

To compute Z scores, click on *Analyze* in the menu bar, then on *Descriptive Statistics*, and finally on *Descriptives*. . . . This will bring up the *Descriptives* dialog box shown in Figure 7-2.

In the dialog box, click on the one variable and then on the ▶ to move it to the *Variable(s)*: area. Now select *Save Standardized values as variables* and then on *OK* to create a new variable in the Data Editor with Z scores. In addition to being presented with the usual descriptive statistics, you have created a new variable titled *Z*Recall in the Data Editor, as shown in Figure 7-3.

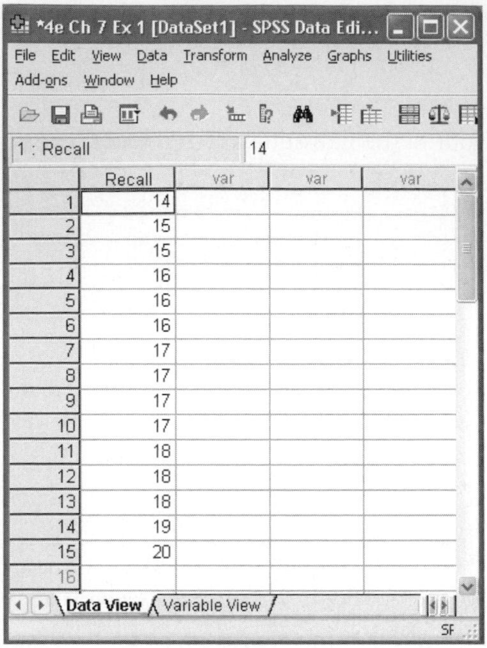

Fig 7-1 Recall test scores.

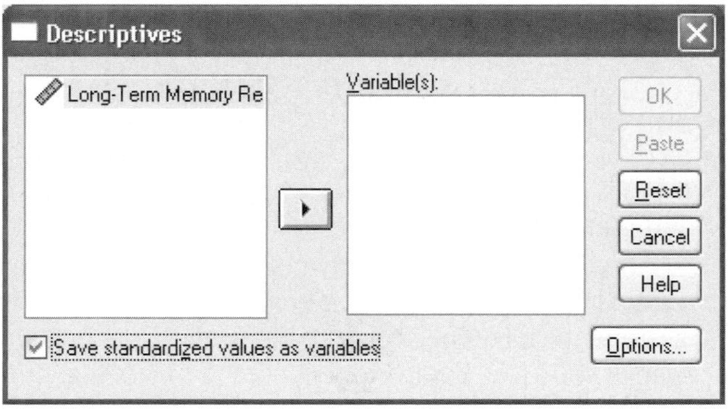

Fig 7-2 Descriptives dialog box.

Let's say Zach obtained a score of 19 on the recall test; for him, then, $X = 19$. As you can see in the ZRecall column, his Z score is 1.34. Jasper's raw score (X) was 14, in contrast, which corresponds to a Z score of -1.79. Notice that the sign in front of the Z score tells you whether the individual's score was above or below the mean. In addition, whenever a person's raw score is equal

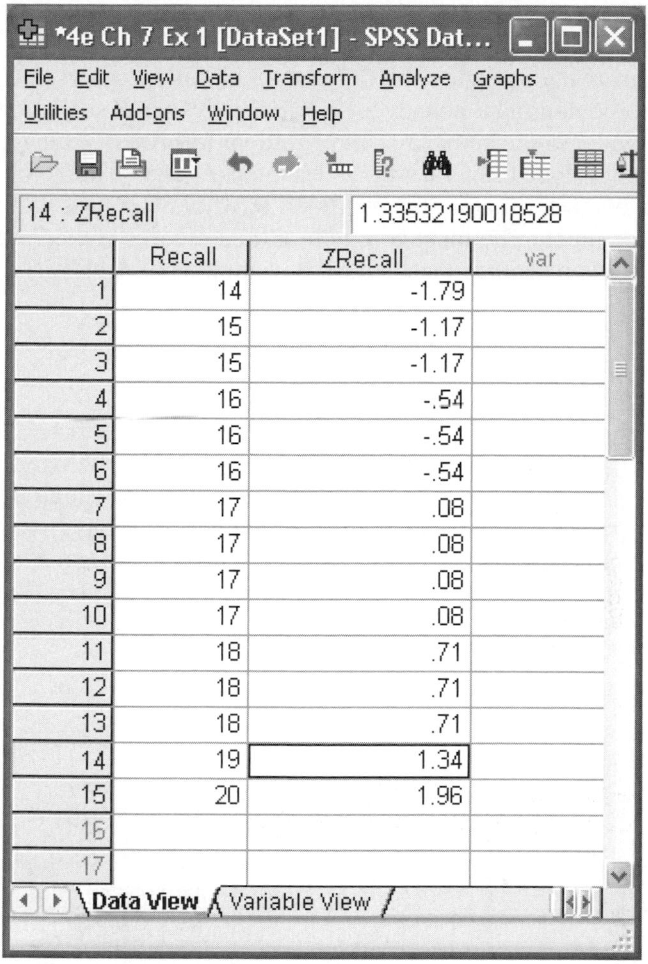

	Recall	ZRecall	var
1	14	-1.79	
2	15	-1.17	
3	15	-1.17	
4	16	-.54	
5	16	-.54	
6	16	-.54	
7	17	.08	
8	17	.08	
9	17	.08	
10	17	.08	
11	18	.71	
12	18	.71	
13	18	.71	
14	19	1.34	
15	20	1.96	
16			
17			

Fig 7-3 *Z* scores.

to the mean, his or her *Z* score equals zero. The mean for these data is 16.87. Although nobody obtained a score exactly at the mean, a score of 17 is slightly above average and equates to a *Z* score of .08, very near zero. All scores that fall at the mean have a *Z* score of zero. If the score had been 1 standard deviation above the mean, the *Z* score would have been +1.0; if it had been 1 standard deviation below the mean, the *Z* score would have been −1.0.

To summarize, the *Z* score tells you if the raw score was above the mean (the *Z* score is positive) or if the raw score was below the mean (the *Z* score is negative), and it tells you how many standard deviations the raw score was above or below the mean.

OTHER STANDARD SCORES

Many people are uncomfortable with negative numbers; others don't like using decimals. Some don't like negatives or decimals! Since Z scores often involve both, these folks would rather not have to deal with them. Our mathematical friends have developed several ways to transform Z scores into other measures that are always positive and often rounded to whole numbers without distorting things too much. The most common of these scores is the T score, which always has a mean of 50 and a standard deviation of 10. We'll discuss T scores next.

T Scores

A number of published personality inventories report test results in a manner similar to that employed by the California Psychological Inventory (CPI). An example of a CPI profile is shown in Figure 7-4. Along the top of the profile, you can see a number of scales designated with letters naming the psychological characteristics measured by the CPI, such as Sc (self-control) and To (tolerance). Under each scale name is a column of numbers. These are raw

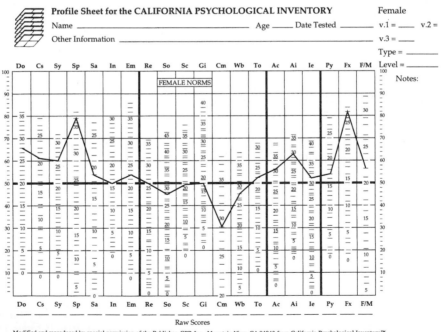

Fig 7-4 Example profile sheet for the California Psychological Inventory.

scores. More interesting to us are the numbers on the left- and right-hand edges. These are the standard score equivalents of the raw scores.

The standard score utilized by the CPI is a T score. The formula for T scores is

$$T = 10(Z) + 50$$

The Z in this formula is the Z score you've just learned to compute.

SPSS makes the task of transforming variables from Z scores to other standard scores quite simple. To convert raw scores to T scores, you must first create Z scores. Once you have done that, click on *Transform* in the menu bar and then on *Compute* to bring up the *Compute Variable* dialog box shown in Figure 7-5.

To create T scores, first type in the name of the new variable, which we will call *T*score, in the *Target Variable*: box in the upper left-hand corner of the dialog box. After this, we need to insert the formula for calculating T scores in the *Numeric Expression*: box. Here we type in the formula for the T score, being careful to use the name of the new Z score variable we created in SPSS. Now click *OK* to transform our Z scores into a new variable with T scores, as shown in Figure 7-6.

As you can see in this figure, if a person's raw score is near the mean of the group, then his or her T score is near 50; raw scores that are below the mean are below 50; and raw scores that are above the mean are above 50.

Let's look at the sample CPI profile again. Notice that the score on the Sc scale has a T score equivalent of almost exactly 50. This T score lets you know

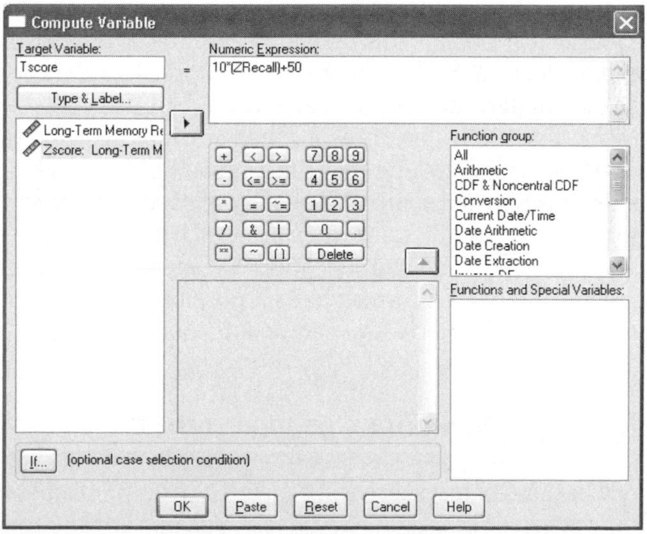

Fig 7-5 Compute variable dialog box.

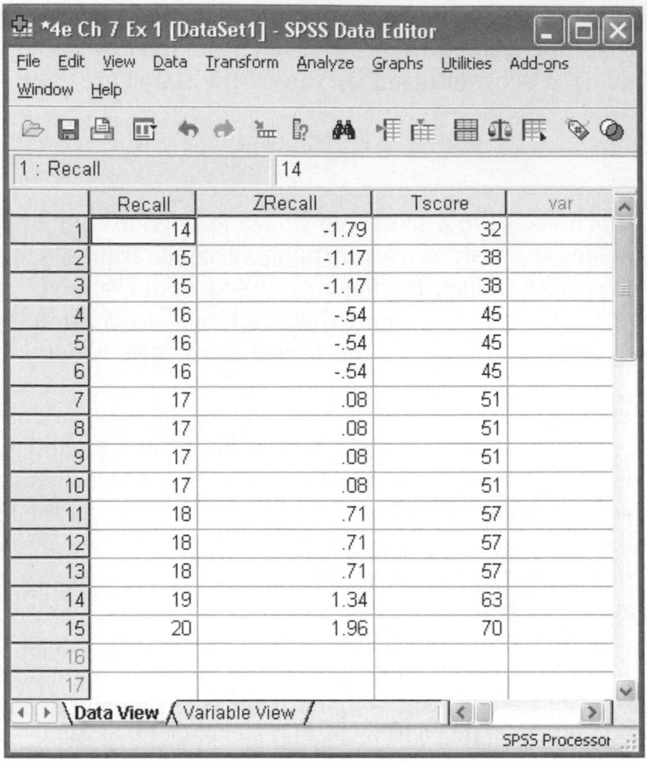

	Recall	ZRecall	Tscore	var
1	14	-1.79	32	
2	15	-1.17	38	
3	15	-1.17	38	
4	16	-.54	45	
5	16	-.54	45	
6	16	-.54	45	
7	17	.08	51	
8	17	.08	51	
9	17	.08	51	
10	17	.08	51	
11	18	.71	57	
12	18	.71	57	
13	18	.71	57	
14	19	1.34	63	
15	20	1.96	70	
16				
17				

Fig 7-6 T scores.

that, relative to the norm group, this person's score was at or near the mean. On the other hand, look at the Sp (social presence) score. The T score is 80, so we know that it is 3 standard deviations above the mean. Very few people in the norm group score that high—the graph in Chapter 6 indicates that well under 1% are up there—so we can conclude that this person had an extremely elevated Sp score. Finally, imagine that someone had a score of 20 on both Sp and Py. Even though their raw scores are the same, they are well below the mean on Sp (their T score would be just under 40) and well above the mean on Py ($T = 61$).

Converting scores to T scores makes it possible to compare them meaningfully. But it doesn't end there—we can do lots more!

CONVERTING STANDARD SCORES TO PERCENTILES

The choice of what kind of test score will be used by a test publisher is somewhat arbitrary. Some types of scores are relatively easy for anyone to understand, whereas others can be really understood only by those who are sophisticated statistically (like you). My own preference of test-score type is the percentile. Percentile rank tells us exactly where a person stands relative to the norm group, without any need for further translation. A percentile rank of 50 corresponds to a score that is exactly in the middle of the norm group—at the median; a percentile rank of 30, in contrast, corresponds to a score that is at the point where 30% of the remaining scores are below it and 70% above it; and a percentile rank of 95 means that 95% of the remaining scores are below it and only 5% of the norm group scores are above it.

Figure 7-7 will help you to see the relationship between standard scores (T's, Z's, IQ scores), percentiles, and the normal curve. If all those numbers look a little threatening to you, don't worry; just take it slowly and it will make perfect sense. With a ruler or other straightedge to guide you, you can use the figure to make a rough conversion from one kind of score to another. For now, though, let's just look at the figure. As you can see, this figure displays several different kinds of standard scores, including Z scores, percentiles, T scores, and Wechsler IQ scores.

To use Figure 7-7 to move among different kinds of scores, you first need to convert a raw score to a Z score. Having done that, you can easily go to any of the other scores. For example, say a score was 1 standard deviation below the mean. This is equivalent to the following:

Z score of -1
T score of 40
Percentile rank of about 16 (15.87)
Wechsler IQ of 85

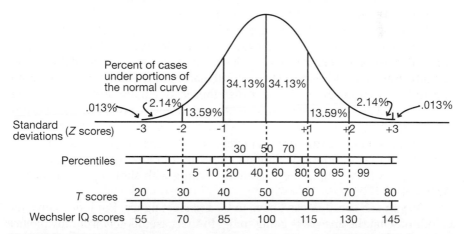

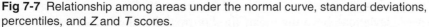

Fig 7-7 Relationship among areas under the normal curve, standard deviations, percentiles, and Z and T scores.

Similarly, a raw score that is 2 standard deviations above the mean is equivalent to the following:

Z score of $+2$

T score of 70

Percentile rank of about 98 (97.72)

Wechsler IQ of 130

Make sense? Not complicated stuff once you get the hang of it.

Chances are that it will be helpful at some time in your work to translate standard scores into approximate percentiles. For example, if you know that a person's MMPI Depression score is 70, and if you know that MMPI scores are T scores, then you also know that he or she scored higher than almost 98% of the norm group on that scale. Similarly, if he or she earned an IQ of 85, you will know that the score is below average and at approximately the 16th percentile.

What is the percentile equivalent of someone with a Z score of $+.5$? You can answer that question by looking down from the Z score scale to the percentile score on Figure 7-7 and making a rough approximation (a Z score of $+.5$ is about the 69th percentile). A more precise answer can be obtained by consulting Appendix B, which presents proportions of area under the standard normal curve.

Don't be nervous; I'll tell you how to use Appendix B. First, though, remember what you already know: If scores are distributed normally, the mean equals the median. Therefore, the mean is equal to the 50th percentile. Recall also (you can check it on Figure 7-7) that a raw score equal to the mean has a Z score equal to zero. Putting these facts together, you can see that $Z = 0 = 50$th percentile. You may also recall from Chapter 6 that a score that is 1 standard deviation above the mean is higher than 84.13% of the norm group, which is another way of saying that a Z score of $+1$ is at the 84th percentile.

Now let's see how to use Appendix B. Look at the following example. Suppose you gave a statistics test to a large group of people, scored their tests, and computed the mean and standard deviation of the raw scores. Assume that the population of scores was distributed normally and that $\overline{X} = 45$, $SD_x = 10$. Justin had a raw score of 58, which corresponds to a Z score of $+1.3$. What is his percentile rank?

At this point, I always draw a picture of a normal (approximately) curve, mark in 1 and 2 standard deviations above and below the mean, and put a check mark where I think the score I'm working with should go, as shown in Figure 7-8.

As you can see, it doesn't have to be a perfect figure. But it does give me an idea of what the answer is going to be. In this case, I know from my picture that the percentile is more than 84 and less than 98. Do you know why? One standard deviation above the mean (in a normal distribution, of course) is at

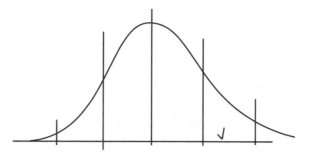

Fig 7-8 Example picture of an approximately normal curve.

the 84th percentile; 2 above the mean is at the 98th percentile. Our score is in between those two benchmarks, probably somewhere in the high 80s. Knowing that helps me to avoid dumb mistakes like reading from the wrong column of the table. If this step helps you too, do it. If not, don't.

Now look in the Z columns of Appendix B until you find the Z score you just obtained ($Z = +1.3$), and write down the first number to the right (in our example, .4032). This number indicates that between the mean and a Z score of $+1.3$, you will find a proportion of .4032, or 40.32% of the whole distribution (to convert proportions to percents, move the decimal two places to the right). You know that 50% of the cases in a normal curve fall below the mean, so a Z score of $+1.3$ is as high or higher than $.50 + .4032 = .9032$, or 90.32% of the cases. Justin's raw score of 58, therefore, corresponds to a percentile rank of 90.32. That's pretty close to my guess of "high 80s"! A more accurate picture of what we just did is shown in Figure 7-9.

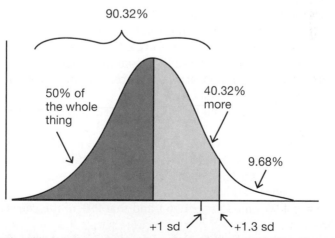

Fig 7-9 Relationship between a raw score of 58, the corresponding Z score of $+1.30$, and the percentile rank of 90.32.

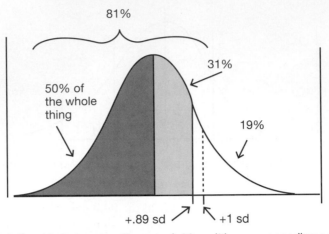

Fig 7-10 Relationship between a Z score of .89 and the corresponding percentile rank of 81.

Another example: In Appendix B, locate a Z score of +.89. The table indicates that 31.33% of the cases fall between the mean ($Z = 0$) and $Z = +.89$, and that 18.67% of the cases fall above +.89. As in the previous problem, the percentile equivalent of $Z = +.89$ is $.50 + .3133 = .8133$, or 81% a percentile rank of approximately 81. See Figure 7-10.

Remember that the standard normal curve is symmetrical. Thus, even though Appendix B shows areas above the mean, the areas below the mean are identical. For example, the percentage of cases between the mean ($Z = 0$) and $Z = -.74$ is about 27%. What is the corresponding percentile rank? Appendix B indicates that beyond $Z = .74$, there are 23% of the cases (third column). Therefore, the percentile rank is 23. See Figure 7-11.

I find that when I sketch a distribution, and use it to get the sense of what I'm looking for, problems like this are easy. When I don't make a sketch, I very often get confused. That's why I recommend that—unless you're very good at this indeed—you always draw a picture. Enough advice, now—back to business!

What percent of cases fall between $Z = +1$ and $Z = -1$? Appendix B indicates that 34.13% fall between the mean and $Z = +1$. Again, since the standard normal curve is symmetrical, there are also 34.13% between the mean and $Z = -1$. Therefore, between $Z = -1$ and $Z = +1$, there will be $34.13\% + 34.13\% = 68.26\%$ of the cases.

Verify for yourself that 95% of the cases fall between $Z = \pm1.96$ (i.e., between $Z = +1.96$ and $Z = -1.96$) and that 99% of the cases lie between $Z = \pm2.58$.

You can use Appendix B to go backwards to find out what Z score corresponds to a given percentile. What Z score marks the point where two-thirds of

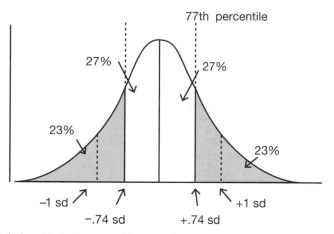

77th percentile

27%

27%

23%

23%

−1 sd

−.74 sd

+.74 sd

+1 sd

Fig 7-11 Relationship between a Z score of −.74 and the corresponding percentile rank of 23.

the scores are below, and one-third above? Well, that would be at the 66.67th percentile. Before you go to Appendix B and look in the second column for 66.67, and get all frustrated because you can't find it, draw a picture, as shown in Figure 7-12.

I've shaded in about two-thirds of this distribution just to see what's going on. Below the mean is 50%; the lighter shaded area above the mean is 16.7%, because the whole shaded part is 66.67%. That's the number I have to find in the table, in the middle column. (The little diagrams at the top of the columns in Appendix B show you what part of the curve is being described in that column.) In this case, we find the number closest to .1667 (.1664) in the far-right column of the first page of Appendix B; reading over to the left, we see that this value corresponds to a Z score of +.43.

Working in the lower half of the distribution is just a bit more complicated, but using a sketch makes it easy. What Z score corresponds to a percentile rank of 30? Picture first: See Figure 7-13.

Since the curve is symmetrical (or it would be if I could draw), you can flip it over and work with its mirror image: See Figure 7-14.

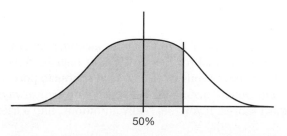

50%

Fig 7-12 Relationship between Z score and the 66.67th percentile.

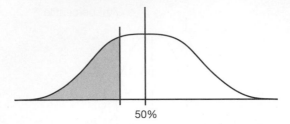

50%

Fig 7-13 Relationship between *Z* score and the 30th percentile.

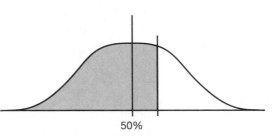

50%

Fig 7-14 Another way to look at the *Z* score–30th percentile relationship.

Thirty percent of the distribution was in the shaded tail of the first picture, so 30% is in the upper unshaded tail of the second, leaving 20% in the lower part of the top half. Looking up that 20%, we find that it corresponds to a *Z* of .52—that is, +.52 standard deviation above the mean. What we want is the *Z* score that is .52 standard deviation below the mean—hey, we know how to do that; that's just $Z = -.52$! If you draw the picture, and use common sense to figure out what you're looking for, you should zip right through these kinds of problems.

A WORD OF WARNING

Remember that all the material in this chapter assumes that the scores you are interpreting are from normal (or nearly normal) distributions. If the distribution is skewed, then conversions from standard scores to percentiles using the normal curve will not be accurate. However, you can always convert raw scores to standard scores. If you feel really ambitious, find a friend and explain why this is so.

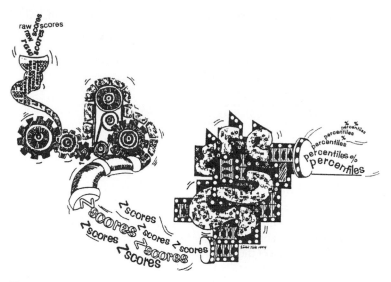

Fig 7-15 Pictorial transformation of raw scores into standard scores.

PROBLEMS

1. The following is a hypothetical set of depression scale scores from a group of ten students being seen at a college counseling center:

	Depress	var	var	var	var
1	16				
2	18				
3	25				
4	46				
5	47				
6	50				
7	25				
8	19				
9	49				
10	48				
11					

*4e Ch 7 Prob 1 [DataSet1] - SPSS Data Editor

File Edit View Data Transform Analyze Graphs Utilities Add-ons Window Help

8 :

Data View / Variable View /

SPSS Process

Find the Z and T scores for these groups. Why is it not a good idea to use Appendix B to compute percentiles for these scores?

2. SAT scores are normally distributed, with a mean of 500 and *SD* of 100. Find the *Z*, *T*, and percentile equivalents of the following scores: 500, 510, 450, 460, 650, and 660.

3. Draw a diagram that shows the relationship of a raw score of 20 to *Z*, *T*, and percentile in a distribution with the following:

   (a) $\bar{X} = 20$, $SD_x = 5$

   (b) $\bar{X} = 40$, $SD_x = 7$

   (c) $\bar{X} = 15$, $SD_x = 4$

ANSWERS TO PROBLEMS

1. Since the scores are not normally distributed, the table of values for a normal curve can't be used to convert these scores to percentiles. (Remember: you can use Variable View to select the number of decimal places to use.)

2.
500 (raw score)	0 (*Z* score)	50 (*T* score)	50th percentile
510 (raw score)	+.1 (*Z* score)	51 (*T* score)	54th percentile
450 (raw score)	−.5 (*Z* score)	45 (*T* score)	31st percentile
460 (raw score)	−.4 (*Z* score)	46 (*T* score)	34th percentile
650 (raw score)	+1.5 (*Z* score)	65 (*T* score)	93rd percentile
660 (raw score)	+1.6 (*Z* score)	66 (*T* score)	95th percentile

3. (a)

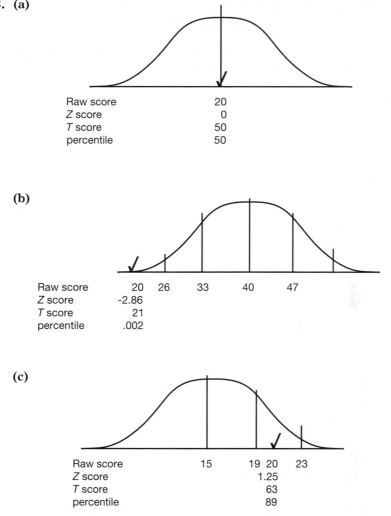

Raw score	20
Z score	0
T score	50
percentile	50

(b)

Raw score	20	26	33	40	47
Z score	-2.86
T score	21
percentile	.002

(c)

Raw score	15	19 20	23
Z score		1.25
T score		63
percentile		89

Correlation and Regression

Right now, I imagine you're feeling pretty good. And you should feel good! We've come a long way. Up to this point, we've focused on describing single variables one at a time. In the preceding chapters, you've learned how to summarize data in the form of frequency distributions and display them graphically. You've learned how to use descriptive statistics to describe the level and spread of a set of scores. And you've learned all about the normal curve and how to use standard scores and percentiles to describe the proportions of scores in different areas of the normal curve. That's a lot! Good job. In this section, you'll learn how to determine whether two variables are related and, if so, how they are related. We'll also find out how to make predictions from one variable to another. In Chapter 8, we discuss the methods of correlation that describe the empirical relationship between two variables. In Chapter 9, we cover how to make predictions with regression equations and how to get an idea of how much error to expect when making those predictions. Let the fun begin!

8

....

Correlation Coefficients

- Correlation Coefficients
- Pearson Product-Moment Correlation Coefficient
- Interpreting Correlation Coefficients
- Creating a Scatterplot
- Other Methods of Correlation
- Spearman Rank Correlation Coefficient
- Problems

> A statistician can have his head in an oven and his feet in ice, and he will say that on average he feels fine.

Up to now, we've been dealing with one variable at a time. In this chapter, we will discuss how to examine the relationship between two variables. Using the methods of correlation, we will be able to answer important questions such as: Are achievement test scores related to grade-point averages? Is counselor empathy related to counseling outcome? Is student toenail length related to success in graduate school? Hey, inquiring minds want to know.

The relationship between two variables can be depicted graphically by means of a scatter diagram. Suppose a group of students took an aptitude test. We'll use a scatter diagram to look at the relationship between Z scores and T scores on this test. We designate their Z scores as the X variable (the X variable is always plotted on the horizontal axis, or "X-axis" of a scattergram) and their T scores will be the Y variable (plotted on the vertical, or "Y-axis"). Each student has two scores. We find a given student's position on the graph by drawing an invisible line out into the center of the graph from the vertical-axis value (the Y value, in this case, the student's T score) and drawing another invisible line up into the graph from the X-axis value (the student's Z score). Put an X where the two lines cross, and you've plotted that student's location.

As you can see in Figure 8-1, all the X's representing the relationship between Z scores and T scores fall on a straight line. When two variables have this kind of relationship, we say that they are perfectly correlated. This means that if we know the value of something on one of the variables, we can figure

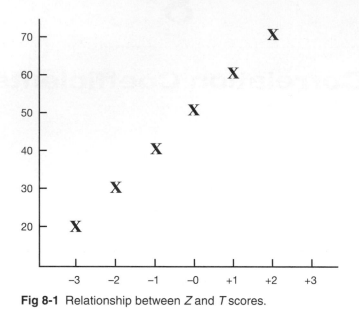

Fig 8-1 Relationship between Z and T scores.

out exactly what its value is on the other. If we know that someone has a T score of 50, the Z score has to be zero. Every time.

CORRELATION COEFFICIENTS

Graphs that show relationships like this in picture form are useful, but often we need a numerical way of expressing the relationship between two variables. In the next few pages, you'll learn how to compute a correlation coefficient. Here are some facts about correlation coefficients:

1. The methods of correlation describe the relationship between two variables. Most methods of correlation were designed to measure the degree to which two variables are linearly related, that is, how closely the relationship resembles a straight line.
2. The correlation coefficient—symbolized by r—indicates the degree of association between two variables.
3. The values of r can range from -1.00 to $+1.00$.
4. The *sign* of the correlation indicates how two variables are related. Positive values of r indicate that low values on one variable are related to low values on another variable, and vice versa. For example, height and weight are positively correlated. Taller people tend to be heavier than shorter people, on average. Negative values of r indicate an inverse

relationship between two variables. Low values on one variable are related to high values on another variable, and vice versa. Intelligence and reaction time, for instance, are negatively related. More intelligent people tend to have lower reaction times on certain laboratory tasks, on average, than people who aren't as intelligent.

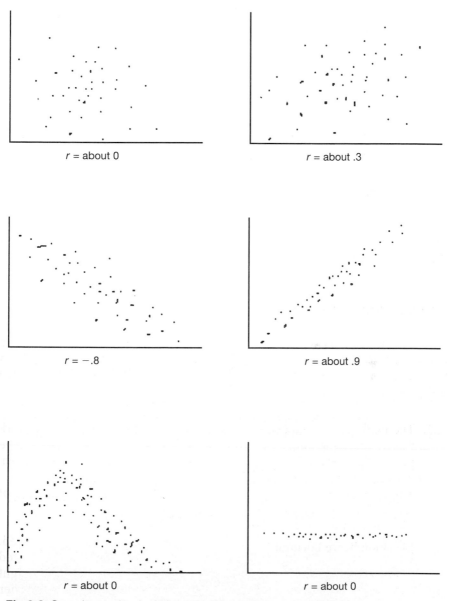

Fig 8-2 Sample graphs depicting various degrees of correlation.

5. When $r = 0$, no relationship between the two variables exists. They are uncorrelated.
6. The absolute value of r (when you ignore the plus or minus sign) indicates how closely two variables are related. Values of r close to -1.00 or $+1.00$ reflect a strong relationship between two variables; values near zero indicate a weak relationship.

Scatterplots and correlation coefficients are related. Figure 8-2 shows example scatterplots depicting various correlation coefficients. As you can see here, correlation coefficients are a quick and easy way to express how and how well two variables are related or go together. The greater the correlation coefficient is, the smaller the scatter around a hypothetical straight line drawn through the middle of the scatterplot. Perfect correlations of -1.00 and $+1.00$ fall on a straight line. High correlations, such as those around $+.90$ or $-.80$, tend to have minimal scatter; whereas lower correlations, such as those around $+.30$, tend to have much more scatter. Also shown in this figure are three ways in which the relationship between two variables can be zero. For each, the values of one variable are totally unrelated to values of the other variable. Keep in mind that a correlation of zero means no *linear* relationship at all. Two variables could be related in a *curvilinear* manner and still have a correlation of zero (as you can see in the lower-left scatterplot in Figure 8-2).

PEARSON PRODUCT-MOMENT CORRELATION COEFFICIENT (r_{xy})

Many measures of correlation have been developed by statisticians for different purposes. The most commonly used measure of correlation is the *Pearson product-moment correlation coefficient*, designated by r_{xy}. The Pearson r_{xy} is used to describe the linear relationship between two quantitative variables. You'll commonly see two formulas for the Pearson r_{xy} in statistics texts—the deviation score formula and the computational formula. The deviation score formula makes more intuitive sense about what is going on when you calculate a Pearson r_{xy}, and the computational formula is a little easier to calculate by hand. Fortunately, due to the advent of statistical software programs like SPSS, few people now do these calculations by hand. So I'll just show you the deviation score formula. Relax. Smile. You have SPSS.

Deviation Score Formula

$$r_{xy} = \frac{Cov_{xy}}{SD_x SD_y}$$

Where: $Cov_{xy} = \dfrac{\Sigma(X - \bar{X})(Y - \bar{Y})}{N - 1}$ or $\dfrac{\Sigma xy}{N - 1}$

$x = (X - \bar{X})$, the deviation scores for X

$y = (Y - \bar{Y})$, the deviation scores for Y

$\Sigma xy =$ the sum of the products of the paired deviation scores

$N =$ number of pairs of scores

$SD_x =$ standard deviation of X

$SD_y =$ standard deviation of Y

If you'll look closely at the formula for the Pearson r_{xy}, you'll notice that many of the values and computations are the same as in calculating the mean and variance. You might want to go back and review the last part of Chapter 5 now if that's still a little fuzzy. The only new term in this formula is the sum of the products of x and y (Σxy). As you can see from the formula, the Pearson r_{xy} is really a kind of average of the deviation scores for two variables. Let's walk through an example of how to calculate the Pearson r_{xy} using SPSS.

Figure 8-3 displays some hypothetical data on variables X and Y for five individuals.

To calculate the correlation between these variables, click on *Analyze* in the menu bar, then on the *Correlate* menu item, and finally on *Bivariate. . . .* This opens the *Bivariate Correlations* dialog box shown in Figure 8-4.

Dialog boxes are probably getting familiar by now. This one is no different. To examine the correlation between variables X and Y, click on each variable and then on the ▶ in the middle of the dialog box to move them to the

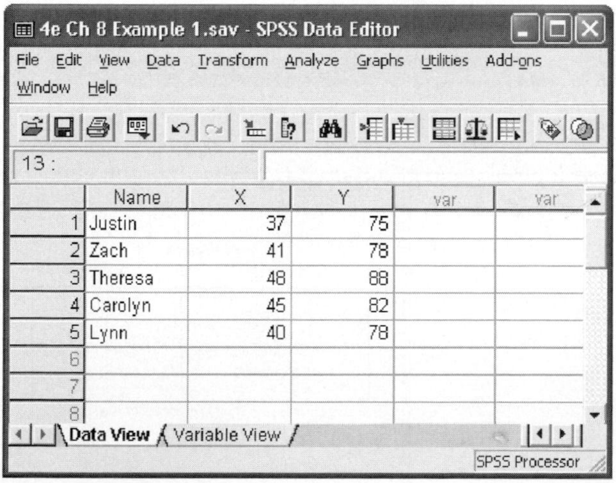

Fig 8-3 Hypothetical data on variables X and Y.

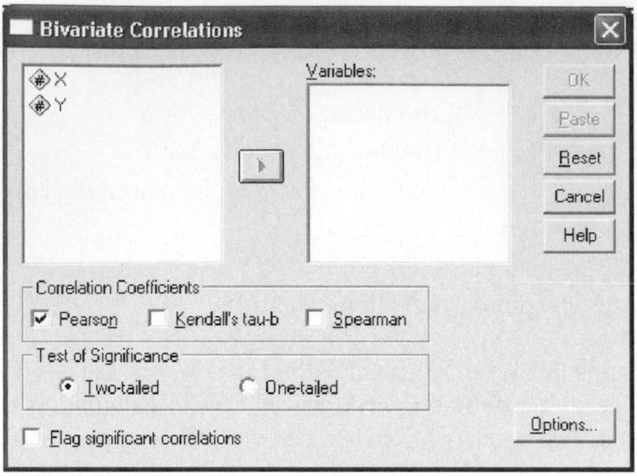

Fig 8-4 Bivariate correlations dialog box.

Variables(s): box. In the *Correlations Coefficients* area, select the box next to *Pearson* to indicate that you want to calculate a Pearson r_{xy}, and then click *OK*. Results of this analysis are shown in the Viewer in Figure 8-5.

Because we have two variables, X and Y, results of this analysis show a 2×2 table (2 rows and 2 columns). The correlation between variables X and Y is $+.977$. The correlation of a variable with itself, of course, is 1.00. The table also shows number (N) of cases, or pairs of scores, that were used in the calculation. The table also contains results for something called *Sig. (2-tailed)*. These are the results of a test of statistical significance. Since we haven't talked

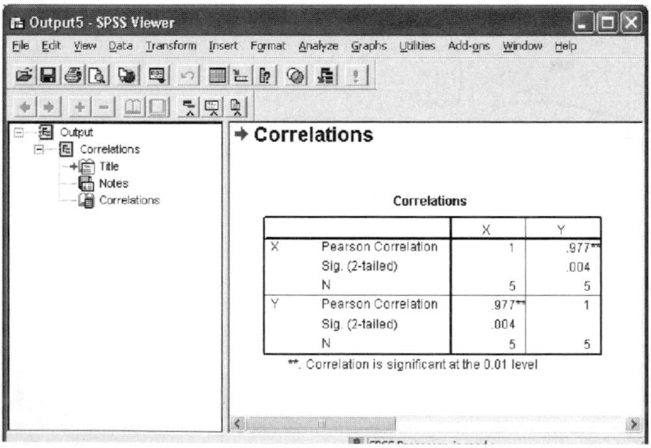

Fig 8-5 Correlation analysis results.

about statistical significance tests, just ignore this for now. We'll cover this in Section III.

INTERPRETING CORRELATION COEFFICIENTS

In our example, we obtained a correlation of +.977 between X and Y. What does this mean? There are a number of possible interpretations. You will recall that a perfect correlation between two variables would result in r_{xy} = +1.00 or −1.00; whereas if there were no correlation at all between X and Y, then r_{xy} = 0. Since the correlation of +.977 is positive and near +1.00, we can say that variables X and Y are highly related, but that the relationship is not perfect. This means that, on average, high scores on X are related to high scores on Y and vice versa.

Another common way to interpret r_{xy} is to calculate what is known as the *coefficient of determination*. The coefficient of determination is simply the square of the correlation coefficient, r_{xy}, or r_{xy}^2. The coefficient of determination is interpreted as the percent of variance in one variable that is "accounted for" or "explained" by knowing the value of the other variable. In other words, when two variables are correlated, the r_{xy}^2 tells us the proportion of variability in one variable that is accounted for by knowing another variable. In our example, the obtained r_{xy} = +.977 and r_{xy}^2 = .95. The percentage of variance in Y explained by X is equal to $r_{xy}^2 \times 100$. Thus, we could say that 95% of the differences in Y are related to differences in X. For example, if we obtained a r_{xy} of −.60 between measures of self-esteem and bullying—meaning that on average high self-esteem is related to less bullying and vice versa —squaring that value would give you .36 (I know that it seems wrong when you see that the square of a number is smaller than the number being squared, but it's true. Try it.) The obtained square of the correlation coefficient (in our example, .36) indicates that 36% of the variability in bullying is "accounted for" or "explained by" the variability in self-esteem. "Why is it that some people achieve more in school than others?" some might ask. If intelligence is related to academic achievement, and if the r_{xy} between intelligence and academic achievement were r_{xy} = .50, then 25% (squaring .50 and converting it to a percentage) of the variability in achievement among people would be "accounted for" by intelligence, the other 75% being independent of intelligence. Another way of saying the same thing is that variables X and Y have 25% of their variability "in common."

It is important not to confuse correlation with causation. In our example above, r_{xy} = .977, the two variables are highly related. But that doesn't prove that X causes Y, any more than it proves that Y causes X. The correlation between the number of counselors and the number of alcoholics in a state may be positive, for example, but that does not mean that one causes the other. Two variables may be correlated with each other due to the common influence of

a third variable. On the other hand, if two variables do not correlate with each other, one variable cannot be the cause of the other. If there is no difference in counseling outcome between, for example, "warm" and "cold" counselors, then "warmth" cannot be a factor determining counseling outcome. Thus, the correlational approach to research can also help us rule out variables that are not important for our theory or practice. This point is so important that it bears repeating: If two variables are correlated, it does not mean that one causes the other. However, if they are not correlated, one cannot be the cause of the other.

CREATING A SCATTERPLOT

To create a scatterplot, click on *Graphs* in the menu bar, then on the *Scatter/Dot . . .* to bring up the *Scatter/Dot* dialog box in Figure 8-6.

In this dialog box, click on *Simple Scatter* and then *Define* to open the *Simple Scatterplot* dialog box shown in Figure 8-7.

Here, click on variable X and then on the ▶ next to the *X Axis:* area to move the variable to that area; then click on variable Y and then on the ▶ next to the *Y Axis:* area to move the variable to that area. Now click *OK* to create the scatterplot shown in Figure 8-8.

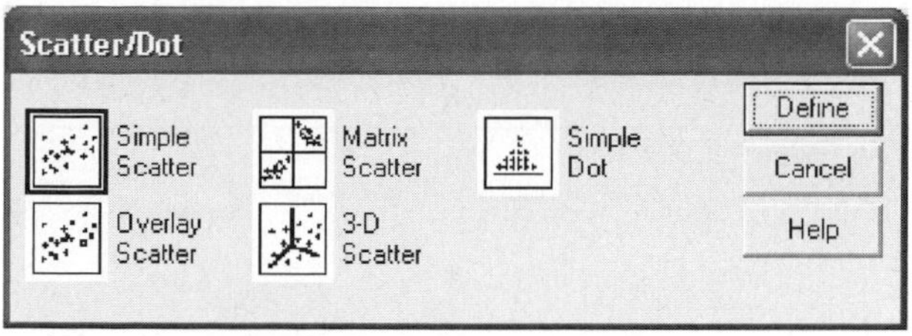

Fig 8-6 Scatter/Dot dialog box.

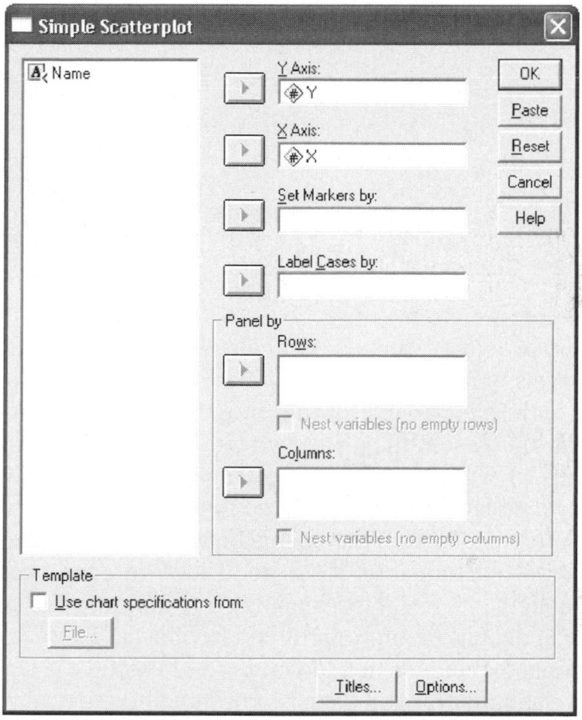

Fig 8-7 Simple scatterplot dialog box.

As you can see in this figure, the points in the scatterplot fall nearly on a straight line, reflecting the very high r_{xy} of .977. That's all there is to it!

OTHER METHODS OF CORRELATION

The Pearson r_{xy} is a measure of the linear relationship between two variables. Some variables are related to each other in a curvilinear fashion, however. For example, the relationship between anxiety and some kinds of performance is

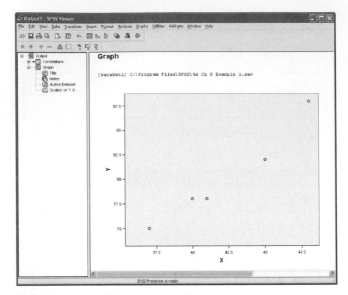

Fig 8-8 Scatterplot of example data.

such that moderate levels of anxiety facilitate performance, whereas extremely high and low levels of anxiety interfere with it. A scatterplot showing the relationship between the two variables would show the dots distributing themselves along a curved line resembling an inverted U (curvilinear), rather than a straight line (linear). The Pearson r_{xy} is not an appropriate measure of relationship between two variables that are related to each other in a curvilinear fashion. If you are interested in the relationship between two variables measured on a large number of participants, it's a good idea to construct a scatterplot for 20 or so participants before calculating r_{xy}, just to get a general idea of whether the relationship (if any) might be curvilinear. There are a number of other correlational techniques in addition to the Pearson r_{xy}. One of these, the Spearman correlation for ranked data, is presented here.

SPEARMAN RANK CORRELATION COEFFICIENT

The Spearman rank correlation (r_s), also called Spearman's *rho*, can be used with variables that are *monotonically* related. A monotonic relationship is one in which the values of one variable tend to increase when the values of the other variable increase, and vice versa, but not necessarily linearly. In other words, if the relationship between two variables is not linear, then the Pearson r_{xy} is inappropriate and another kind of correlation should be used, such as the Spearman r_s. In addition, the r_s can be used with ranked data.

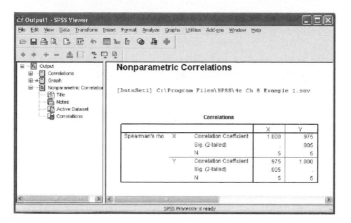

Fig 8-9 Spearman r_s results.

When scores on variables X and Y are ranked, the following formula for the Spearman r_s may be used:

$$r_s = \frac{Cov_{xy}}{SD_x SD_y} = \frac{\frac{\Sigma xy}{N-1}}{SD_x SD_y}$$

Where: X, Y are ranks

In SPSS, we don't need to go to the trouble of actually ranking data. SPSS will do it for us. To calculate the Spearman r_s between our example variables, then, simply click on the *Analyze* menu on the menu bar, move your mouse over the *Correlate* menu item and click on *Bivariate.* . . . This opens the *Bivariate Correlations* dialog box shown above in Figure 8-4. Click on variables X and Y and then on the ▶ in the middle of the dialog box to move them to the *Variables(s):* box. In the *Correlations Coefficients* area, select the box next to *Spearman* to indicate that you want to calculate a Spearman r_s and click *OK*. Results of this analysis are shown in the Viewer in Figure 8-9.

As you can see in this figure, results for the Spearman r_s are almost identical to those for the Pearson r_{xy}.

PROBLEMS

1. Given the following data, what is the Pearson r_{xy} and coefficient of determination between:
   (a) IQ scores and anxiety test scores?
   (b) IQ scores and statistics exam scores?
   (c) anxiety test scores and statistics exam scores?

2. A researcher is interested in determining the relationship between the popularity of high school students and academic rank in class. Using these ranked data, (a) calculate Spearman r_s and the coefficient of determination; and (b) create a scatterplot for these data.

ANSWERS TO PROBLEMS

1. (a) $r_{xy} = -.606$; $r_{xy}^2 = .367$, or 36.7%.
    (b) $r_{xy} = +.880$; $r_{xy}^2 = .774$, or 77.4%.
    (c) $r_{xy} = -.661$; $r_{xy}^2 = .437$, or 43.7%.

2. (a) $r_s = -.353$; $r_s^2 = .125$, or 12.5%.

(b)

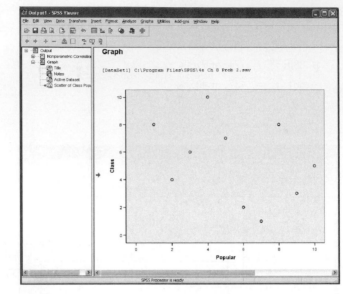

9

Linear Regression

- Standard Error of the Estimate
- Problems

> Statistical Analysis: Mysterious, sometimes bizarre, manipulations performed upon the collected data of an experiment in order to obscure the fact that the results have no generalizable meaning for humanity. Commonly, computers are used, lending an additional aura of unreality to the proceedings.

Statisticians and researchers and teachers and weather forecasters and all sorts of other folks are interested in making predictions. A prediction, in this sense, is simply a best guess as to the value of something. We try to make our predictions so that, over the long run, the difference between the value we predict and the actual value (what the thing we are trying to predict really turns out to be) is as small as possible.

If we have no additional information, the best guess that we can make about a value is to predict that it will equal the mean of the distribution it comes from. If I want to predict John's score on a test, and I know that the average score students like John have received on that test is 50, my best prediction of John's score is 50. Over time, across lots and lots of predictions, always predicting that somebody's score will equal the mean of the distribution will give me the best chance of coming close to a correct answer. Making predictions with no extra information is not very useful, however. If the weatherperson always made exactly the same forecast (variable clouds, possibility of showers) day after day, people would quickly stop paying attention. Most often, though, we do have additional information, and we use that information to improve the accuracy of our prediction.

You know from Chapter 8 that when there is a positive correlation between two variables, X and Y, those persons who score high on the X variable also tend to score high on the Y variable, and those who score low on X tend to score low on Y.[1] For example, suppose you know that there is a positive correlation between shoe size and weight. Given a person's shoe size, you can improve your guess about his weight. If you knew that John's shoe size was 12, you would predict that he weighs more than Jim, whose shoe size is 7. You can increase the accuracy of predictions of this sort considerably if you use what statisticians call a regression equation. If two variables are correlated, it is possible to predict with greater than chance accuracy the score on one variable from another with the use of a regression equation. The higher the correlation

CLOSE TO HOME   JOHN McPHERSON

© 1996 John McPherson/Dist. by Universal Press Syndicate

Unfortunately, Brad had neglected to stretch his brain before taking the big algebra midterm.

is between two variables, the better the prediction. If the r_{xy} is +1.00 or −1.00, then prediction is perfect. In this case, knowledge of a person's standing on variable X tells us exactly where he or she stands on variable Y. If the r_{xy} is zero, then prediction is no better than chance.

The following is an intuitive form of the regression equation:

$$\hat{Y} = \bar{Y} + b(X - \bar{X})$$

The predicted value of $Y(\hat{Y})$ is equal to the mean of the Y distribution $(\bar{Y})$—no surprise here—plus something that has to do with the value of X. Part of that "something" is easy—it's how far X is from its own mean $(\bar{X})$. That makes sense: Having a shoe size that is way above the mean would lead us to predict that weight would also be above the mean, and vice versa. That takes care of the $(X - \bar{X})$ part. But what's the "b" stand for? Well, b is a term that adjusts our prediction so that we take into account differences in the means and the standard deviations of the two distributions with which we're working, as well as the fact that they aren't per- fectly correlated. In our example, if IQ scores and achievement test scores had a correlation of +1.00, and if they had the same mean and the same standard devia- tion, the value of b would turn out to be 1 and we could just ignore it.

It isn't necessary to go into the details of where b comes from or why it works to make our predictions—all we need to know for now is how to calcu- late it. Fortunately, it's fairly easy to calculate:

$$b = r_{xy}\left(\frac{SD_y}{SD_x}\right)$$

So, to make a prediction about Y, based on knowing the corresponding value of X, all you need to know is the correlation between X and Y (that's the Pearson r_{xy} we computed in Chapter 8), the means and standard deviations of both dis- tributions, and the value of X.

SPSS, of course, eliminates the need to calculate any of these statistics by hand. SPSS, however, uses a slightly different formula, although both produce the same results. Here is the regression equation used in SPSS:

$$\hat{Y} = B_{slope} X + B_{constant}$$

In this formula, the Y scores $(\hat{Y})$ are predicted from a slope weight (B_{slope}) for the predictor variable (X) and an additive constant $(B_{constant})$. Let's do an exam- ple and see just how easy this is to do.

Suppose you wanted to predict performance on the job (Y) from a test of knowledge about the job (X) with the data in Figure 9-1. For this example, then, Job Knowledge is the independent variable and Job Performance is the dependent variable.

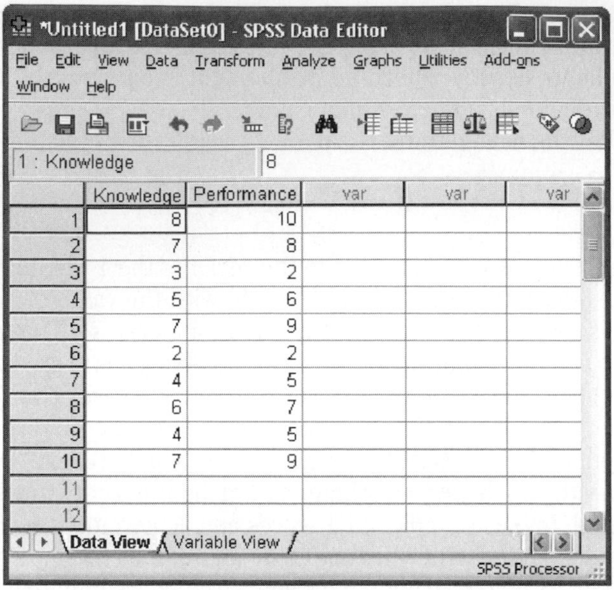

Fig 9-1 Example regression data.

To run the regression analysis, click on *Analyze* in the menu bar, then on the *Regression* menu item, and finally click on *Linear*. . . . This opens the *Linear Regression* dialog box shown in Figure 9-2.

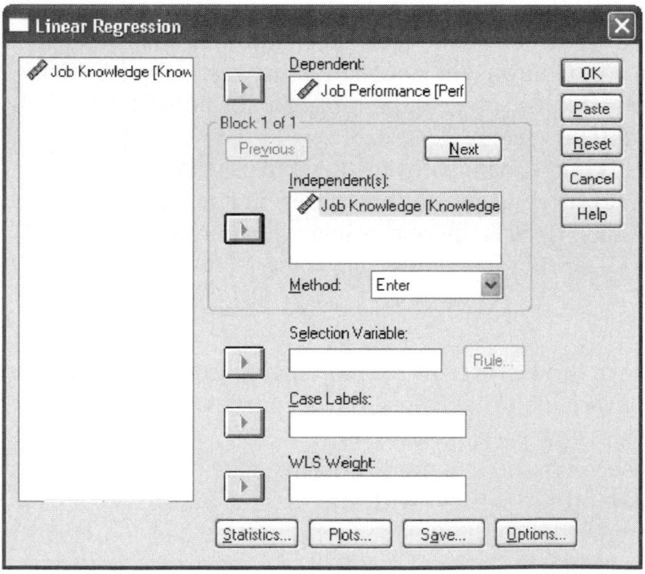

Fig 9-2 Linear regression dialog box.

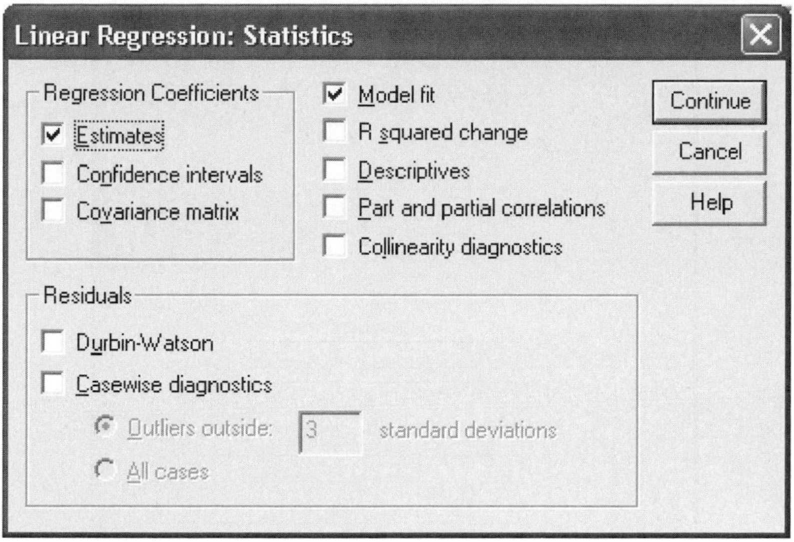

Fig 9-3 Linear regression: Statistics dialog box.

Click on our Job Knowledge and then on the ▶ next to the *Independent(s)*: area, because this variable is our independent variable, the variable we are using as a predictor of job performance; now click on Job Performance and then on the ▶ next to the *Dependent:* area, because this variable is our dependent variable, the variable we want to predict. Make sure that *Enter* is shown in the *Method*: area. After you have done this, click on the *Statistics . . .* box to bring up the *Linear Regression: Statistics* dialog box shown in Figure 9-3.

In this dialog box make sure that *Estimates* and *Model Fit* are checked and then click *Continue*. Once you are back at the *Linear Regression* dialog box, click *OK* to run the analysis. Figure 9-4 displays the output for the linear regression analysis.

The results that we are interested in are the slope weight (B_{slope}) and the additive constant ($B_{constant}$), both of which can be found in the *Unstandardized Coefficients* section of the box labeled *Coefficients* in the output. As you can see here, the B_{slope} = 1.388 and the $B_{constant}$ = −1.055. Now that these values have been calculated, we can complete our regression equation for predicting job performance from job knowledge:

$$\hat{Y} = B_{slope}X + B_{constant}$$

$$\hat{Y} = (1.388)X + (-1.055)$$

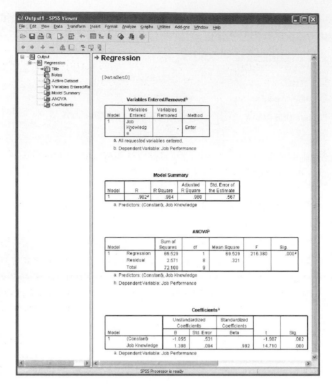

Fig 9-4 Linear regression analysis results.

With the regression equation completed, we can predict for any score on X the predicted value on Y. For example, for $X = 7$, we have

$$\hat{Y} = (1.388)X + (-1.055) = (1.388 \times 7) + (-1.055) = 8.66$$

Thus, for a person with a job knowledge score of 7, our best guess of actual job performance would be 8.66.

STANDARD ERROR OF THE ESTIMATE

Schools and employers use a variety of measures to predict how well people will do in their settings. Although such predictions usually are more accurate than those made by nonmathematical methods (guesstimation), there is some danger in using the results of predictions in a mechanical way. Suppose, for example, that the University of Michifornia uses an admissions test to predict the expected GPA of incoming students. Over the years, it has been established that the mean score on the Test of Admissions (X) is 50.0 (SD = 10) and the mean Frosh GPA (Y) is 2.40 (SD = 0.60). The correlation between these two variables is +.70. After conducting a linear regression analysis, they found a regression or prediction equation of

$$\hat{Y} = (.042)X + (.302)$$

Denny, a 12th grader, took the admissions test and got a score of 34. To predict his GPA using linear regression, we calculate:

$$\hat{Y} = (.042)X + (.302) = (.042 \times 34) + .302 = 1.73$$

Since his predicted GPA of 1.73 is less than a "C" average, it might be tempting to conclude that Denny is not "college material." Before jumping to that conclusion, however, it is wise to remember that predictions about individual performance aren't perfectly accurate. Not everyone who has a predicted GPA of 1.73 achieves exactly that: Some do better than predicted (overachievers?), and others do worse (underachievers?).

In fact, if you took 100 persons, all of whom had a predicted GPA of 1.73, their actual GPAs would vary considerably—and they would form a distribution. If you could look into a possible future and actually see how well these students did, you could compute a mean and standard deviation of the distribution of their GPAs. Theoretically, the scores would be distributed normally and would have a mean equal to the predicted GPA of 1.73 and a standard deviation of the predicted GPA of

$$\sigma_{est} = (SD_y\sqrt{1-r_{xy}^2})\sqrt{\frac{N-1}{N-2}}$$

Where: σ_{est} is the standard error of the estimate
N is the number of pairs of observations
r_{xy}^2 is squared correlation between the X and Y variables

Now this is a very important idea. We're talking about a group of people, selected out of probably thousands who took the college admissions test— selected because that test predicted that they would get a 1.73 GPA. But they don't all get a GPA of 1.73: Some do better than predicted and some not so well. Their GPAs will tend to cluster around the predicted value, however. If there were enough of them (it would probably take more than 100 in order to smooth out the curve), the distribution would be normal in form, would have a mean equal to the predicted value, and would have a standard deviation of

$$\sigma_{est} = (SD_y\sqrt{1-r_{xy}^2})\sqrt{\frac{N-1}{N-2}}$$
$$= .60\sqrt{1-.70^2} = .60\sqrt{1-.49} = .60\sqrt{.51} = .60(.71) = .43$$

Notice that the term $\dfrac{\sqrt{N-1}}{\sqrt{N-2}}$ is reduced to 1 and drops out of the equation because the sample is so large. This term doesn't matter much when your sample is large; it can matter a lot with a small sample.

You will recall from Chapter 6 that approximately 68% of the scores in a normal distribution fall between ±1 standard deviation from the mean. In this example, the standard deviation of our distribution of GPA scores from those low-test-score students is .43. This value, .43, is known as the standard error of the estimate (σ_{est}). So, we conclude that 68% of the GPAs in this distribution can be expected to fall between 1.73 − .43, and 1.73 + .43, or between 1.30 and 2.16. In other words, 68% of the scores fall between ±1 standard error of the estimate from the mean.

In Figure 9-5, I've shaded the area that represents GPAs of 2.00 and above. If you define someone who graduates from college with a "C" average as "college material," then you can see that a fairly good percentage of those who were predicted to have a 1.73 GPA (less than a "C" average) actually would do all right (i.e., have a GPA higher than 2.00). In fact, using the table in Appendix B, you can figure out exactly how many students are likely to fall into this category:

1. A GPA of 2.00 is .27 grade points above the mean (2.00 − 1.73).
2. That's .27/.43 of a standard deviation, or .63 standard deviation above the mean.
3. Going to the table, we find that .26 of the distribution lies above a Z score of .63: More than a quarter of the GPAs in this distribution will be 2.00 or better.

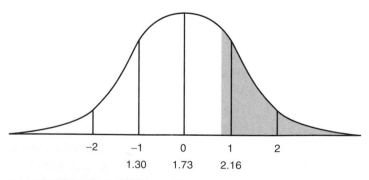

Fig 9-5 Proportion of GPAs of 2.0 and above.

4. About 26% of the total population of students would be mislabeled if we used a cutting off point of exactly 1.73.

It's all very logical if you just take it one step at a time.

Using the standard error of the estimate, we can report predictions in a way that tells others how accurate the prediction is. If we were to say that a student who scores 34 on the admissions test will have a GPA of 1.73±.43, we would be right about two-thirds of the time. (If that confuses you, better go back to Chapter 6 and read about proportions of scores under the normal curve.) And that means, of course, that we'd still be wrong about one-third of the time—about 16% of folks getting that score on the test would do better than a 2.16 GPA, and about 16% would do worse than a 1.30 GPA.

Do you see now why σ_{est} is called the standard error of the estimate? It gives us a way of estimating how much error there will be in using a particular prediction formula—a particular regression equation.

Look at the formula one more time:

$$\sigma_{est} = SD_y\sqrt{(1 - r_{xy}^2)}$$

Notice that if $r_{xy} = 1.00$, then $\sigma_{est} = 0$, which is another way of saying that if the correlation between X and Y were perfect, there would be no errors in predicting performance on Y from our predictor X. Unfortunately (or fortunately?), there are virtually no perfect correlations between predictor and predicted variables when we try to measure human beings. We are not able to predict individual human performance with anything near perfection at this time—and, between you and me, nor do I believe we ever will.

Fortunately, calculating the σ_{est} in SPSS is very easy to do. In fact, we've already done it by selecting the *Model Fit* option in the *Linear Regression* dialog box. As you can see in the output displayed in Figure 9–4, the σ_{est} can be found in the box titled *Model Summary* under *Std Error of the Estimate*. For our example, the σ_{est} is .567.

PROBLEMS

1. A developmental psychologist has following data (in months) on when babies and their mothers spoke their first sentences:

Create a regression equation for predicting when babies born in 2005 will most likely speak their first sentences based on knowledge of when their mothers spoke their first sentences.

2. If we know that Theresa said her first sentence at 10.8 months, what is our best guess as to when her child, Jake, will speak his first sentence?

3. What is the standard error of estimate, σ_{est}, for this regression equation?

4. Find the range of ages at which baby Jake can be expected to first string words together to form a sentence with an approximately 68% chance of being correct.

ANSWERS TO PROBLEMS

1. $\hat{Y} = B_{slope}X + B_{constant} = .984X + .484$

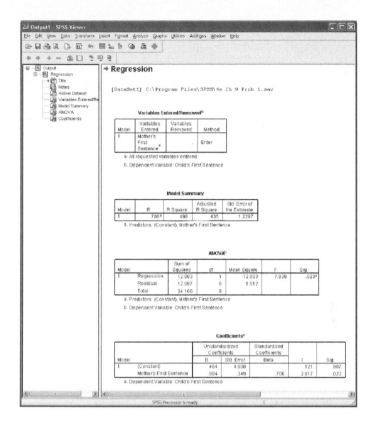

2. $\hat{Y} = .984X + .484 = .984(10.8) + .484 = 11.1$ months

3. $\sigma_{est} = 1.23$

4. Between 9.87 months and 12.33 months.

ENDNOTE

1. It's important to remember that regression equations don't imply anything about causality—even though they may appear to do so. Regressions equations, like correlations, simply reflect the fact that two variables are related.

Inferential Statistics

Three sections down and one to go. In this section we discuss inferential statistics. Inferential statistics involve the use of statistics to infer things about larger populations by studying smaller samples of those larger populations. A word of warning: Of all the chapters in this book, Chapter 10 is probably the most important and possibly the most difficult. This chapter discusses the basic concepts underlying the use of statistics to make inferences. The difficulty of this chapter has nothing to do with the complexity of the math involved—because there is very little math in this chapter—and everything to do with the abstractness of the ideas involved in using statistics to test hypotheses. If you grasp the main concepts in this chapter, the rest of the book will be a piece of cake. So take your time reading Chapter 10. Chapter 11 discusses how to use the basic concepts covered in the preceding chapter to examine differences between the means of two groups with the *t* test. Chapter 12 covers analysis of variance (ANOVA), a method for analyzing differences between the means of two or more groups. In Chapter 13, the concept of nonparametric statistics is introduced and the most common nonparametric test called chi-square is discussed. Chapter 14 consists of the postscript in which we review how far you have come and congratulate you on a job truly well done.

10

....

Introduction to Inferential Statistics

- Probability
- The Sampling Experiment
- Sample Values (Statistics) and Population Values (Parameters)
- The Null Hypothesis
- Type I and Type II Errors
- Statistical Significance and Type I Error
- Problems

> The larger the sample size (n) the more confident you can be that your sample mean is a good representation of the population mean. In other words, the "n" justifies the means.

We've all had the experience of flipping on the evening news and getting the weather forecast. In Chapter 9, we talked about the kind of forecast the weatherperson would have to make if he or she had no information at all except the most common weather situation that occurs in that location. For Oregon, that would likely be "tomorrow it will rain." But the weatherperson doesn't say that; the forecast is more like, "Tomorrow there is an 80% chance of showers." What, exactly, does that "80% chance of showers" mean? Will it rain 80% of the time tomorrow? Or, maybe, 80% of us will get rained on? Or, we'll have 80% of a full rainstorm sometime during the day? Those are pretty silly interpretations; we all know that an "80% chance of rain" means that it's very likely to rain—but then again, maybe it won't. More precisely, it means that in the past, with conditions like this, there was rain the next day 80% of the time. And, one time out of five, the rain didn't happen.

"There's an 80% chance of rain tomorrow" is a probability statement. Such statements tell us how likely—how probable—it is that a particular event will occur. Probability statements are statistical statements, but, unlike the kind of statistics we've been looking at so far, they go beyond simply describing a set of data that we have in hand. They represent ways of describing and predicting what we don't know, on the basis of current data.

As we'll see in later chapters of this book, probability statements play an important role in research. The whole notion of statistical significance rests

Mrs. Zackford liked her students to feel the full shock effect of a pop quiz.

http://www.ucomics.com/closetohome/2005/11/08/

on probability statements. "Significance" here has a very specific meaning, but I'll wait to give you the formal definition later, when it will make more sense. For now, I just want to introduce you to some of the concepts of simple probability.

PROBABILITY

Almost every explanation of probability starts with a description of the process of flipping coins, and this will be no exception. I'm tired of "heads" and "tails," though. My coin will be a Canadian $1 coin, generally known as a "loony" because it has a picture of a loon on one side—the other side depicts the Queen. So, instead of "heads" and "tails," we'll be talking about "loons" and "Queens."

If I flip my coin 600 times, and if it's a fair coin (equally likely to come up either way), how many of those times should I expect it to come up loons? Right, half the time—300 loons. And 300 Queens. We would say that the probability of getting a loon is 50%, or .50. What if I'm throwing a die instead of flipping a coin? A die has six sides, and if it's a fair die, each side is equally likely to come up. So I would expect that my 600 throws would yield 100 ones, 100 twos, 100 threes, and so on. Will I always get the same numbers if I repeat the coin-tossing experiment over and over again? Nope, there'll be some randomness in the data I generate. I might get 305 loons and only 295 Queens in one set of coin flips, or 289 loons and 311 Queens in another. And there would be similar fluctuations if I repeated the die-throwing experiment over and over again. But, over many, many sets of coin flips or die throws, the deviations would even out. The more often I repeated the experiment, the more closely my total data would approximate the predicted percentages. Over an infinite number of experiments, the numbers of loons and Queens, or of ones-twos-threes-fours-fives-sixes, would be exactly as predicted. Probability statements don't tell us exactly what will happen, but they are our best guess about what will happen. Making our predictions about events based on known probabilities will, over the long run, result in less error than making them any other way.

There are a couple of things that I'd like you to notice about those two situations, flipping the coin or throwing a die. First, each throw comes out only one way—it's an all-or-nothing situation. You can't flip a coin and have it come out half loon and half Queen (unless it stands on edge, and I won't let that happen); you can't throw a die and get 2.35 as a result. Second, each possible outcome is known. There are only two ways that my coin can land; and there are exactly six possible outcomes when we throw a die. So we know exactly what those possible outcomes are.

Okay, I'm going to flip that loony again. This time, though, I want you to tell me the probability of getting either a loon *or* a Queen. Silly question, you say—I'll always get one or the other. Exactly right—the probability of getting either a loon or a Queen is 100%, or 1.00. The probability of getting a loon is .50, the probability of getting a Queen is .50, and the probability of getting a loon or a Queen is 1.00. Do you see a rule coming? When two mutually exclusive[1] outcomes have known probabilities, the probability of getting either the one or the other in a given experiment is the sum of their individual probabilities. What's the probability of our die throw yielding either a five or a six? If you added $\frac{1}{6}$ and $\frac{1}{6}$ and got $\frac{2}{6}$, or $\frac{1}{3}$, or .33, you've got the idea!

Let's do just one more loony experiment (yes, I really said that), and then move on to something more interesting. This time we're going to flip two coins. How many possible outcomes are there, and what is the probability of each? Before you answer, let me warn you that this is a trick question. The most obvious answer is that there are three possible outcomes: two loons, or two Queens, or a loon and a Queen. So far, so good. But, if you go on to say that since the coins are all fair coins, and loons and Queens are equally likely, the probability of each of those three outcomes is .33, then you've fallen for the trick. Look at the possible outcomes more closely:

Outcome	Coin A	Coin B
1	Loon	Loon
2	Loon	Queen
3	Queen	Loon
4	Queen	Queen

Even though it looks, on the surface, as if only three different things can happen, there are actually four possible outcomes. And each is equally likely. Knowing that, we can easily determine the probabilities of tossing 2 loons, 2 Queens, or a Queen and a loon. The probability of two loons is 1 in 4, or .25. The probability of 2 Queens is the same, .25. And the probability of a loon and a Queen is the sum of the probabilities of the two ways of getting that outcome: Loon on Coin A and Queen on Coin B has a probability of .25; Queen on Coin A and loon on Coin B has a probability of .25; the probability of getting exactly one loon and one Queen when we toss two coins is .25 + .25, or .50. Putting it in standard symbols:

$$\rho_{L,Q} = .50$$

THE SAMPLING EXPERIMENT

Most of the time, in the kinds of research that call for statistical procedures, we aren't able to look at the total population in which we are interested. Gathering data on everyone in a population—such as all the children in special education in the United States—is almost always too time-consuming or expensive. We therefore have to use relatively small samples to represent the much larger populations. If I do an experiment that involves testing the reading skills of third graders, I might work with a sample of 50 or 100 children in the third grade. But I'm not really interested in just those 50 or 100 children; I want to say something that will hold true for all third graders. Agriculturists who test the usefulness of fertilizers or pesticides want to make predictions about how those fertilizers or pesticides will work for all the crops on which they might be used, not just describe the small sample they've tested. And looking at the effects of a new vaccine on a sample of 20 volunteers would be pretty useless unless we expected to generalize those results to lots and lots of other folks. Sampling, and predicting outcomes on the basis of those samples, is a fundamental idea in research. That's why we need to understand the concept of probability in sampling.

We'll start with some definitions. I've already used the words, and I hope you've understood what they mean, but it's a good idea to get the precise definitions out on the table:

- *Population:* a large (sometimes infinitely large) group about which some information is desired. Examples include:
  All third-grade children in the United States
  All the wheat crops in North America
  Height of all men entering basic training in the armed forces
  All the beans in a 5-gallon jar
- *Sample:* a subset of a population; a smaller group selected from the population. Examples include:
  First-grade children in Mrs. Walton's class at Wiles Elementary School
  Southernmost acre of wheat in each farm in Ayres County
  Height of the first 100 men entering Army training in Florida in January, 2006
  The first 10 beans drawn from a 5-gallon jar of beans
- *Random sample:* a sample selected in such a way that (a) every member of the population from which it is drawn has an equal chance of being selected, and (b) selection of one member has no effect on the selection of any other member.

The randomness of a sample is very, very important, because everything we are going to say about samples and populations only holds (for sure) when the sample is truly random. That's why I put that jar of beans in as an example: If we stir up the beans very thoroughly, and then pull out a sample with our eyes shut, then every bean has an equal chance of ending up in our sample. Jars of beans are favorite tools for statistics teachers, because they do yield random samples.

Our jar of beans, by the way, has two colors of beans in it, red and white. Exactly half of the beans are red and exactly half are white. Okay, stir them up, close your eyes, and pull one out. Don't look at it yet! Before you open your eyes, what color bean do you think you drew? Of course, you can't make a very good guess—since half of the beans are red and half are white, you have an equal chance of drawing a red bean or a white one. Over in Jar 2, 80% of the beans are red. Mix them up, close your eyes, draw one, and guess its color. You'll guess red, of course, and in the long run that guess will be correct 80% of the time. The probability of getting a red bean is equal to the proportion of red beans in the total population. For Jar 1, $p_{Red} = .50$; for Jar 2, $p_{Red} = .80$.

If you drew a random sample of not one bean, but ten beans, from each jar, what would those samples look like? Intuitively, we know the answer: The sample from Jar 1 would have about half red and half white beans; the sample from Jar 2 would have about 80% red beans. Samples tend to resemble the populations from which they are drawn. More specifically, over time, the proportions of the sample (for the characteristic of interest) approximate more and more closely the proportions of that characteristic in the parent population. And the bigger the sample, the more likely it is to resemble the parent population. The more experiments we do with Jar 1, the more the number of red and white beans will tend to equal out. The larger the samples we draw from Jar 2, the closer our samples will come to having exactly 80% red and 20% white beans. If we had a magic jar, with an infinite supply of beans, these statements would be true whether or not we put each sample back into the jar before we took another. This basic characteristic of random samples, that over time and with increasing size they come to resemble the parent population more and more closely, is true for samples taken with or without replacement.

SAMPLE VALUES (STATISTICS) AND POPULATION VALUES (PARAMETERS)

You can see where all this is going, can't you? Since it's often difficult or even impossible to measure an entire population, it makes sense to get a sample from that population and measure it instead. What's the average weight of 10-year-olds? Get a sample of 10-year-olds, and weigh them. How much do North Americans spend on groceries every week? Get 50 or so folks to let you keep track of what they do in the supermarket. What's the effect of second-hand smoke on the water intake of white rats? Put a dozen rats in a cage, blow smoke at them, and measure what they drink. We can find the means and standard deviations of samples like this and use them to estimate the same values in the populations from which they came. Or can we?

There are a couple of problems that must be overcome in order to use sample values (technically, these are called "statistics") to estimate population values (called "parameters"). One has to do with the representativeness of the sample: If we are going to use sample measurements to estimate population values, the sample has to be truly representative of the population. The experiment with the 10-year-olds wouldn't yield very useful data, for instance, if we got the children in our sample from a group of young gymnasts. Kids involved in gymnastics tend to be thinner than the general population of kids. The easiest way to ensure that our sample is representative is to select it randomly.[2] That way, even though it may not be exactly like the parent population, the differences will at least tend to even out over a large number of samples, or as a single sample gets larger. Samples that are truly representative of their parent population are said to be unbiased. Perhaps the most notorious biased sample was taken inadvertently in 1936, when the *Literary Digest* predicted on the basis of its sample of automobile and telephone users that Alf Landon, the Republican candidate, would win the presidential election by a landslide. In 1936, many voters didn't have telephones or cars, and the ones who didn't have them were more likely to be Democrats than Republicans. Of course, Alf Landon didn't win by a landslide; he lost to Franklin D. Roosevelt.

Even with an unbiased sample, however, bias can creep into some estimates of population values. Some statistics—mathematical descriptions—are inherently biased, and some aren't. The two statistics that are most important to us at this point are the mean and the standard deviation. The mean of a sample is an *unbiased* estimate of the mean of the population from which it came. Hmm . . . that's a complicated-sounding sentence. Let's introduce a couple of new symbols to simplify things. The mean of a population is designated by μ, the Greek letter mu. (If we're talking about samples, we use our standard English "$\overline{X}$" for the mean; and if we're talking about populations, we use the Greek "μ." Similarly with the variance and standard deviation: For samples, we use the English "SD^2" for the variance and "SD" for the standard deviation; and for populations we use the Greek lowercase "σ^2" and "σ.") Now, getting back to business—if we take lots and lots of samples from a given population, the means of those samples ($\overline{X}$) will form a distribution that clusters around the mean of the population (μ). What's more, that distribution will be normal—but that's another story, one we'll save for later.

Remember, back when we first talked about the variance, I mentioned in Chapter 5 that you will often see the squared deviations from the mean divided by $N - 1$ instead of by N, in order to get SD^2? Well, now we're ready to talk about why that is. When we compute the variance of a sample, we are actually estimating the variance of the population from which the sample was drawn. If we wanted to find the true "average squared deviation," we would use this formula:

$$SD_x^2 = \frac{\Sigma (X - \overline{X})^2}{N}$$

Using this formula, which involves dividing by N, we would get the variance of the sample itself, but this value would not be an unbiased estimate of the population variance. This value would tend to be smaller than the population variance. The smaller the sample, the greater this error is likely to be. Dividing by $N - 1$ instead of by N is a way of correcting the problem—that is, of correcting for bias in estimating the population value. When the sample is very large, the difference between N and $N - 1$ is negligible; little correction is made, and little is needed. With small samples, the correction is larger. Just as it should be.

Perhaps looking at two extreme situations will help you see how this works. What's the smallest sample you could draw? That's right, just one case, $N = 1$. If we were to find one without correcting for bias, we'd get

$$SD_x^2 = \frac{\Sigma (X - \bar{X})^2}{N} = \frac{\Sigma (0)^2}{1} = 0$$

No matter how large the population variance might be, this "sample" wouldn't reflect any of it at all.

What's the biggest sample you could draw? Right again, one the size of the whole population. And its variance (again, uncorrected) would be exactly the same as SD_x^2. So, we have this situation shown in Figure 10-1.

It turns out, by the way, that the points in between those two check marks don't fall in a straight line—SD^2 and σ^2—aren't linearly related. Instead, SD^2 and σ^2 can be quite different when the sample is small, but the difference levels out relatively quickly, something like that shown in Figure 10-2.

This situation leads to a very handy fact. Since the amount of bias in using SD^2 to estimate σ^2 is proportional to the size of the sample (N), we can

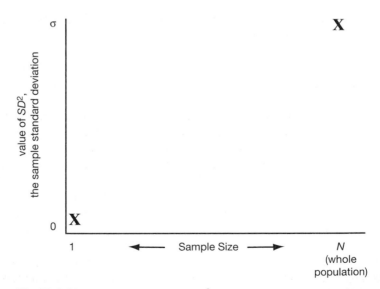

Fig 10-1 Uncorrected values of SD^2 for small and large samples.

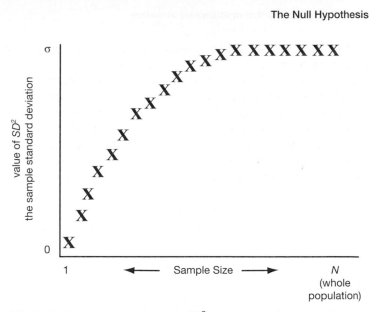

Fig 10-2 Uncorrected values of SD^2 for a range of sample sizes.

correct for that bias by substituting $N - 1$ for N in the formula for the variance:

$$SD_x^2 = \frac{\sum (X - \bar{X})^2}{N - 1}$$

This is the same formula for the variance that we have been using all along, however. So this is nothing new.

THE NULL HYPOTHESIS

Ordinarily, researchers are interested in demonstrating the truth of some hypothesis of interest: that some relationship exists between two variables; that two groups differ in some important way; that Population A is bigger, stronger, smarter, more anxious, and so on than Population B. We need statistical procedures to test such hypotheses. The problem is, though, that it's almost impossible (with most statistical techniques) to demonstrate that something is true. Statistical techniques are much better at demonstrating that a particular hypothesis is false, that it's very unlikely that the hypothesis could really hold up.

So we have an interesting dilemma. We want to show that something is true, but our best tools only know how to show that something is false. The solution is both logical and elegant: State the exact opposite of what we want to demonstrate to be true, disprove that, and what's left—what hasn't been disproved—must be true.

In case you are now thoroughly confused, here's an example: A researcher wants to show that boys, in general, have larger ears than girls. (I know, it's a curious idea, but research grants have been awarded for stranger things.) She knows that her statistical tools can't be used to demonstrate the truth of her hypothesis. So she constructs what's known as a null hypothesis, which takes in every possibility except the one thing she wants to prove: Boys' ears are either smaller than or just the same size as girls' ears. If she can use her statistical techniques to disprove or reject this null hypothesis, then there's only one thing left to believe about boys' and girls' ears—the very thing she wanted to prove in the first place. So, if you can rule out all the gray parts, the only thing left (the thing that must be true) is the white part—exactly what you wanted to prove all along. See Figure 10-3.

The hypothesis that the scientist wants to support or prove is known as the research hypothesis, symbolized H_1; the "everything else" hypothesis is called the null hypothesis and is symbolized as H_0. A primary use of inferential statistics is that of attempting to reject H_0.

Okay, time for an example. Suppose we wanted to compare the math anxiety of male and female graduate students in the United States. In theory, we could administer a math anxiety test to all female and male graduate students in the country, score the tests, and compare the μ's for the two populations. Chances are good, however, that our resources would not allow us to conduct

This whole blob represents all the possible ways the world might be.

And the big gray area represents everything that *isn't* what you'd like to prove.

The little white patch represents the thing you'd like to prove is true.

So, if you can rule out all the gray parts, the only thing left (the thing that must be true) is the white part—exactly what you wanted to prove in the first place.

Fig 10-3 A blob diagram of the world.

such a study. So we decide to use statistics. If we suspect that there is a difference in the two populations, and that's what we want to demonstrate, then our H_0 is that there is no difference. In statistical notation:

$$H_0: \mu_{\text{Females}} = \mu_{\text{Males}}$$

The purpose of our study is to decide whether H_0 is probably true or probably false.

Suppose we drew 50 subjects (using a random sampling method, of course) from each of the two populations, male graduate students and female graduate students, and tested them for math anxiety. Suppose also that the mean of our female sample was 60 and the mean for our male sample was 50:

$$\bar{X}_{\text{Females}} = 60 \qquad \bar{X}_{\text{Males}} = 50$$

Obviously, the mean of our female sample was higher than that for males. But how do we use this fact to justify throwing out—that is, rejecting—the null hypothesis, so as to be left with the hypothesis we are trying to prove?

There are two possible explanations for our observed difference between male and female sample means: (a) There is, in fact, a difference in math anxiety between the male and female population means, that is, $\bar{X}_{\text{Female}} \neq \bar{X}_{\text{Males}}$, and the difference we see between the samples reflects this fact; or (b) there is no appreciable difference in anxiety between the means of the male and female graduate student populations, that is, $\bar{X}_{\text{Female}} = \bar{X}_{\text{Males}}$, and the difference we observe between the sample means is due to chance, or sampling error (this would be analogous to drawing more white beans than red even though there is no difference in the proportion of the two in the jar from which the sample was drawn).

If the null hypothesis really is true, and there really is no difference in math anxiety between male and female graduate students, then the differences we observe between sample means is due to chance. The statistical tests you will study in later chapters, such as the t test in Chapter 11 or analysis of variance (ANOVA) in Chapter 12, will help you to decide if your obtained result is likely to have been due to chance or if there probably is a difference in the two populations. If the result of your study is that the difference you observe is "statistically significant," then you will reject the null hypothesis and conclude that you believe there is a real difference in the two populations. Of course, even after concluding that the difference was real, you would still have to decide whether it was large enough, in the context of your research situation, to have practical usefulness. Researchers call this the "magnitude of effect" question.

TYPE I AND TYPE II ERRORS

Your decision either to reject or not to reject the null hypothesis is subject to error. Because you have not studied all members of both populations and because statistics (such as $\overline{X}$ and SD_x) are subject to sampling error, you can never be completely sure whether H_0 is true or not. In drawing your conclusion about H_0, you can make two kinds of errors, known (cleverly) as Type I and Type II errors. Rejection of a true null hypothesis is a Type I error, and failure to reject a false null hypothesis is a Type II error. Perhaps the following table will help you to understand the difference.

Investigator's Decision	The Real Situation (unknown to the investigator)	
	H_0 Is True	H_0 Is False
Reject H_0	Investigator makes a Type I error	Investigator makes a Correct decision
Do not reject H_0	Investigator makes a Correct decision	Investigator makes a Type II error

If the real situation is that there is no difference in math anxiety between males and females, but you reject H_0, then you have made a Type I error; if you do not reject H_0, then you have made a correct decision. On the other hand, if there is a real difference between the population means of males and females and you reject H_0, you have made a correct decision, but if you fail to reject H_0, you have made a Type II error.

The primary purpose of inferential statistics is to help you to decide whether to reject the null hypothesis and to estimate the probability of a Type I or Type II error when making your decision. Inferential statistics can't tell you for sure whether you've made either a Type I or a Type II error, but they can tell you how likely it is that you have made one.

One last point: Notice that you do not have the option of accepting the null hypothesis. That would amount to using your statistical test to "prove" that the null hypothesis is true, and you can't do that. Your two possible decisions really amount to either: (a) I reject the null hypothesis, and so I believe that there really are differences between the two populations; or (b) I fail to reject the null hypothesis, and I still don't know whether there are important differences or not. For this reason, Type I errors are generally considered to be more serious than Type II errors. Claiming significant results when there really are no differences between the populations is held to be a more serious mistake than saying that you don't know for sure, even when those differences might exist. As you will see, statistical decisions are usually made so as to minimize the likelihood of a Type I error, even at the risk of making lots of Type II errors.

STATISTICAL SIGNIFICANCE AND TYPE I ERROR

Suppose that a colleague of yours actually did the study of math anxiety that we've been talking about and concluded that the difference between male and female sample means was "statistically significant at the .05 (or the 5%) level." This statement would mean that a difference as big as or bigger than what he observed between the sample means could have occurred only 5 times out of 100 by chance alone. Since it could have happened only 5 times out of 100 just by chance, your colleague may be willing to bet that there is a real difference in the populations of male and female graduate students and will reject the null hypothesis.

You must realize, however, that whenever you reject the null hypothesis, you may be making an error. Perhaps the null hypothesis really is true, and this is one of those 5 times out of 100 when, by chance alone, you got this large a difference in sample means. Another way of saying the same thing is that if the null hypothesis were true, 5 times out of 100 you would make a Type I error when you use this decision rule. You would reject the null hypothesis when it was, in fact, true 5% of the time.

You might say, "But, I don't want to make errors! Why can't I use the .01 (or 1%) level of significance instead of the .05 level? That way, I reduce the likelihood of a Type I error to 1 out of 100 times." You can do that, of course, but when you do so, you increase the probability of a Type II error. That's because you reduce the probability of a Type I error by insisting on more stringent conditions for accepting your research hypothesis—that is, you fail to reject H_0 even when H_0 is fairly unlikely. Reducing the probability of a Type I error from .05 to .01 means that you'll fail to reject H_0 even if the odds are 98 out of 100 that it is untrue. In educational and psychological research, it is conventional to set the .05 level of significance as a minimum standard for the rejection of the null hypothesis. Typically, if an obtained result is significant at, say, the .08 level, a researcher will conclude that he or she was unable to reject the null hypothesis or that the results were not statistically significant. Making a Type II error, failing to reject the null hypothesis when it's really not true, is rather like missing your plane at the airport. You didn't do the right thing this time, but you can still catch another plane. Making a Type I error is like getting on the wrong plane—not only did you miss the right one, but now you're headed in the wrong direction! With a result significant at the .08 level, the odds still suggest that the null hypothesis is false (you'd only get this large a difference by chance 8 times in 100). But researchers don't want to get on the wrong airplane; they'd rather make a Type II error and wait for another chance to reject H_0 in some future study.

Remember our discussion about magnitude of effect, when we were looking at correlation coefficients? Just because an observed difference may be significant (i.e., may represent real differences between populations), it is not necessarily useful. We might reject the null hypothesis with regard to differences between male and female math anxiety scores, but if the real difference were only one or two points, with means of 50 and 60, would those differences be worth paying attention to?

Magnitude of effect is often ignored in media reports of medical discoveries. "Using toothpicks linked to the rare disease archeomelanitis!" trumpet the headlines, and the news story goes on to say that researchers have shown that people who use toothpicks are 1.5 times more likely to get archeomelanitis than people who don't. But if you look at the actual numbers, the likelihood of getting archeomelanitis (with no toothpicks) is one in a million, and so using a toothpick increases those odds to $1\frac{1}{2}$ in a million. Are you going to worry about raising your risk from .00001 to .000015? I'm not.

PROBLEMS

1. Answer the following questions:
   (a) What is the difference between $\overline{X}$ and μ? Between SD_x and σ?
   (b) Which of the four preceding symbols represents a value that has been corrected for bias? Why not the others?

2. Which of the following would not be a truly random sample, and why?
   (a) To get a sample of children attending a particular grade school, the researcher numbered all the children on the playground at recess time and used a random-number table to select 50 of them.
   (b) Another researcher, at a different school, got a list of all the families who had kids in that school. She wrote each name on a slip of paper, mixed up the slips, and drew out a name. All kids with that last name went into the sample; she kept this up until she had 50 kids.
   (c) A third researcher took a ruler to the school office, where student files were kept. She used a random-number table to get a number, measured that distance into the files, and the child whose file she was over at that point was selected for the sample. She did this 50 times and selected 50 kids.
   (d) A fourth researcher listed all the kids in the school and numbered the names. He then used a random-number table to pick out 50 names.

3. What is the appropriate H_0 for the following research situation?
   (a) A study to investigate possible differences in academic achievement between right- and left-handed children

4. Assume that each of the following statements is in error: Each describes a researcher's conclusions, but the researcher is mistaken. Indicate whether the error is Type I or Type II.
   (a) "The data indicate that there are significant differences between males and females in their ability to perform Task 1."
   (b) "There are no significant differences between males and females in their ability to perform Task 2."
   (c) "On the basis of our data, we reject the null hypothesis."

5. Answer the following questions:
   (a) Explain, in words, the meaning of the following: "The difference between Group 1 and Group 2 is significant at the .05 level."

(b) When would a researcher be likely to use the .01 level of significance rather than the .05 level? What is the drawback of using the .01 level?

ANSWERS TO PROBLEMS

1. (a) $\bar{X}$ and SD_x are sample values, and μ and σ are population values;

   (b) The value of SD_x is obtained from a formula that includes a correction for bias. If this correction weren't in the formula, the obtained value would tend to be too small. $\bar{X}$ doesn't need such a correction because it is already an unbiased estimate. μ and σ don't need correction, because they are themselves population values.

2. (a) Not random because all members of the population didn't have an equal chance of being included (kids who were ill or stayed inside during recess couldn't be chosen).

   (b) Not random because selection wasn't independent (once a given child was selected, anyone else with that last name was included, too).

   (c) Not random, because kids who had very thick files (which usually means they caused some sort of problem) would have a better chance of being selected).

   (d) (Okay, okay, so it was too easy . . .) Random.

3. (a) $\mu_{Left} = \mu_{Right}$

4. (a) Type I   (b) Type II   (c) Type I

5. (a) If we performed this experiment over and over and if the null hypothesis were true, we could expect to get these results just by chance only 5 times out of 100.

   (b) We use the .01 level when we need to be very sure that we are not making a Type I error. The drawback is that, as we reduce the probability of a Type I error, the likelihood of a Type II error goes up.

ENDNOTES

1. "Mutually exclusive" is a fancy way of saying that you can have one or the other, but not both at the same time.

2. One of the easiest ways to get an (approximately) random sample is by using a random-number table or use SPSS to select a random sample for you. Appendix C shows you how to select a random sample in SPSS.

11

The *t* Test

- The *t* Test for Independent Samples
- Assumptions for Using the *t* Test for Independent Samples
- Formula for the *t* Test for Independent Samples
- Conducting a *t* Test for Independent Samples
- The *t* Test for Dependent (Matched) Samples
- Directional Versus Nondirectional Tests
- Problems

Top Ten Reasons to be a Statistician

1. Estimating parameters is easier than dealing with real life.
2. Statisticians are significant.
3. I always wanted to learn the entire Greek alphabet.
4. The probability a statistician major will get a job is $> .9999$.
5. If I flunk out I can always transfer to engineering.
6. We do it with confidence, frequency, and variability.
7. You never have to be right—only close.
8. We're normal and everyone else is skewed.
9. The regression line looks better than the unemployment line.
10. No one knows what we do so we are always right.

The *t* test is one of the most commonly used statistical tests. Its primary purpose is to determine whether the means of two groups of scores differ to a statistically significant degree. Here's an example: Suppose that you randomly assigned 12 participants to a counseling group (Group 1) and 12 to a waiting-list control group (Group 2) that received no counseling. Suppose also that after those in Group 1 had been counseled, you administered a measure of psychological adjustment to the two groups, with results as shown in Figure 11-1.

As you can see in Figure 11-1, you have two variables in the study. The first variable is *Adjust*, which is the measure of adjustment; and the second variable is a grouping variable, which indicates whether a participant was assigned to the counseling group or the control group. The latter variable, of

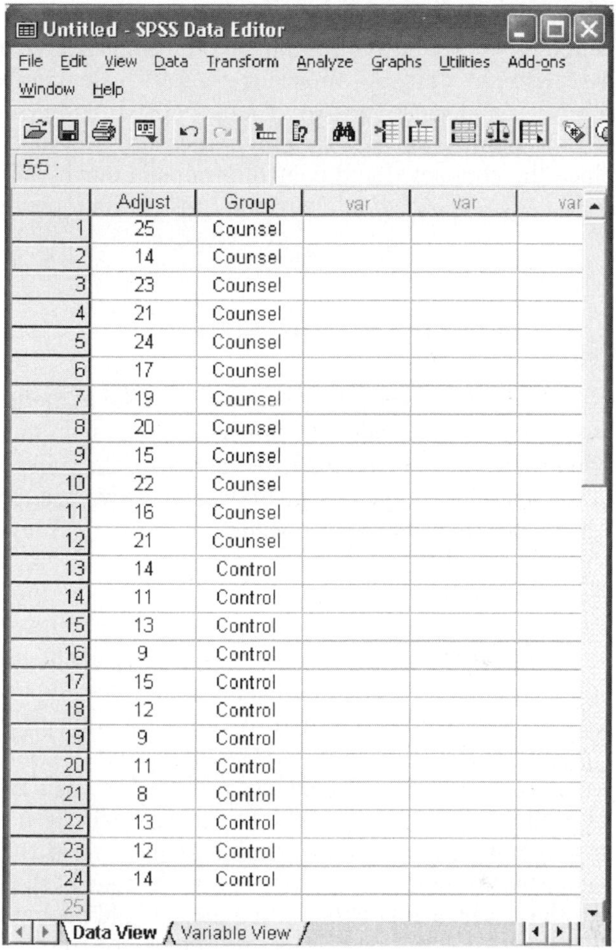

	Adjust	Group	var	var	var
1	25	Counsel			
2	14	Counsel			
3	23	Counsel			
4	21	Counsel			
5	24	Counsel			
6	17	Counsel			
7	19	Counsel			
8	20	Counsel			
9	15	Counsel			
10	22	Counsel			
11	16	Counsel			
12	21	Counsel			
13	14	Control			
14	11	Control			
15	13	Control			
16	9	Control			
17	15	Control			
18	12	Control			
19	9	Control			
20	11	Control			
21	8	Control			
22	13	Control			
23	12	Control			
24	14	Control			
25					

Fig 11-1 Example results.

course, is a string variable (i.e., alphanumeric). For this study, since you want to examine whether the two groups are really different, your null hypothesis is

$$H_0: \mu_1 = \mu_2$$

The null hypothesis states that there is no difference in mean adjustment level between those in the population who receive counseling and those who don't. As you can see, there is a difference in the two sample means, but it may be that this observed difference occurred by chance and there really is no difference in the population means. We need to find out if the difference is statistically significant. If the difference between $\bar{X}_1$ and $\bar{X}_2$ is statistically significant, you will reject the null hypothesis and conclude that there is a difference in the adjustment level between the people who have had counseling and those who have not.

There are two kinds of *t* tests—those for groups whose members are independent of each other and those for two groups whose members are paired in some way (like pretreatment and posttreatment measures, for instance, or pairs of siblings). Since the counseled and control groups in our hypothetical study were not paired in any way, we would use the *t* test for independent samples. Here we go!

THE *t* TEST FOR INDEPENDENT SAMPLES

The *t* test, like most other statistical tests, consists of a set of mathematical procedures that yields a numerical value. In the case of the *t* test, the value that is obtained is called *t*. The larger the absolute value of *t*, the more likely it is to reflect a statistically significant difference between the two groups under comparison. We'll learn how to compute the value of *t* in the next few pages; first, though, let's think about what a *t* test really examines.

A major application of the *t* test for independent samples is found in experimental research like the study in our example. Researchers often draw a sample from one population and randomly assign half of the subjects to an experimental and the other half to a control group, or to some other comparison group (see Appendix C for a method of assigning subjects at random to two or more groups). They then administer some sort of treatment to the experimental group. Since the subjects were assigned to their two groups by chance (i.e., at random), the means of the two groups should not differ from each other at the beginning of the experiment any more than would be expected on the basis of chance alone. If a *t* test were done at the beginning of the experiment, the difference between the means would probably not be statistically significant.[1] After the treatment, the means of the two groups are compared using the *t* test. If the absolute value of *t* is large enough to be statistically significant, the experimenter rejects H_0. Since the two groups have now been shown to differ more than would be expected on the basis of chance alone, and since the only difference between them (that we know of) is the experimental treatment, it is reasonable to conclude that this treatment is responsible for the differences that were observed.

ASSUMPTIONS FOR USING THE *t* TEST FOR
INDEPENDENT SAMPLES

The independent-samples *t* test assumes that you have two independent samples, which means that the subjects for one group were selected independently from those in the second group. That is, the measurements from the two groups aren't paired in any way; a given measurement from Group 1 doesn't "go with" a particular measurement from Group 2. Sometimes you want to do a study in which this is not the case; for paired data you will use the *t* test for dependent (matched) groups.

Also, *t* tests assume that both of the populations being considered are essentially normally distributed. I say "essentially," because a really close fit to the normal distribution isn't necessary. The *t* test is considered "robust" with respect to this assumption—that is, we can violate the assumption without putting too much strain on our findings, especially if we have large samples. If you have only small samples (as in the last example we worked), and if the populations they came from are likely to be quite skewed, then you should not use a *t* test—you should use a nonparametric test. An example of a widely used nonparametric test is chi-square, which we'll discuss in Chapter 13. If the assumption of equal variances is violated, then the results of the statistical test should not be trusted. When you run the independent-samples *t* test procedure SPSS provides you with the results of two different *t* tests: one that assumes equality of variances, and one that does not (Levene's test). If the variances are *not* statistically significantly different, according to the results of Levene's test, you use the results of the *t* test that assumes equal variances; if the variances are statistically significantly different, again according to the results of Levene's test, then you would use the results of the *t* test that does not assume that the variances are equal. Confusing? If it is, just hang on for a bit. It'll be clear after we go through an example.

FORMULA FOR THE *t* TEST FOR INDEPENDENT SAMPLES

The formula for the *t* test looks pretty horrendous at first. When you look closely, you will see that you already learned how to do the computations in earlier chapters. This formula for the *t* test for independent samples can be used with samples of equal and unequal sizes:

$$t = \frac{\bar{X}_1 - \bar{X}_2}{\sqrt{\left[\dfrac{(n_1 - 1)SD_1^2 + (n_2 - 1)SD_2^2}{n_1 + n_2 - 2}\right]\left(\dfrac{1}{n_1} + \dfrac{1}{n_2}\right)}}$$

Where: t = the value of t obtained through your data
n_1, n_2 = the number of participants in Group 1 (n_1)
        and Group 2 (n_2)
SD_1^2, SD_2^2 = the estimates of the variances of the two populations
$\bar{X}_1, \bar{X}_2$ = the means of the two groups

CONDUCTING A *t* TEST FOR INDEPENDENT SAMPLES

To conduct a *t* test for independent samples at the .05 level of significance, click on *Analyze* in the menu bar; then click on *Compare Means*, and finally on the *Independent-Samples T Test*. This will bring up the *Independent-Samples T Test* dialog box shown in Figure 11-2.

Click on the variable, *Adjust*, and then on the ▶ to move it to the *Test Variable(s)*: box; and then click on *Group* and the ▶ next to the *Grouping Variable*: area. Next, click on *Define Groups* to bring up the *Define Groups* dialog box shown in Figure 11-3.

In the Group 1 box type in *Counsel*, and in the Group 2 box type in *Control*. This tells SPSS that you want to compare the means of these two groups on the *Adjust* variable. Now click *Continue* and then *OK* in the *Independent-Samples T Test* dialog box to run the analysis. The results of this analysis are shown in Figure 11-4.

As you can see in this figure, the *Group Statistics* box displays descriptive statistics for the two variables being compared. The results of our *t* test are shown in the box labeled *Independent Samples Test*. The difference between the two means is 8.000 and the *t* is 6.557.

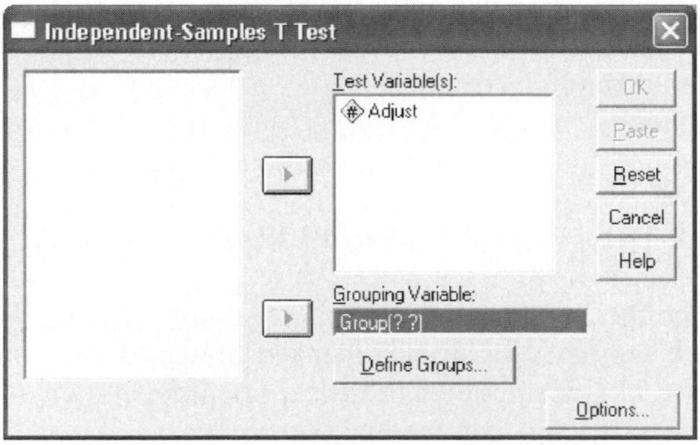

Fig 11-2 Independent-Samples *t* Test dialog box.

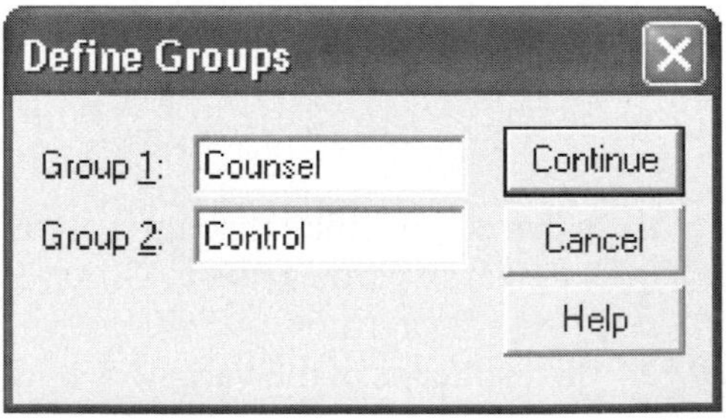

Fig 11-3 Define Groups dialog box.

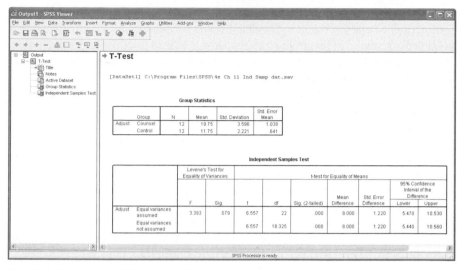

Fig 11-4 Independent-Samples *t* Test results.

"Okay," you might say, "I've done all the computations. Now what does my *t* mean?" Good question. To find out whether your *t* is statistically significant—that is, if it is large enough so that it probably reflects more than chance or random differences between the two samples—simply look at the value in the *Sig. (2-tailed)* column.[2] If the value is less than the level of significance chosen beforehand, then you reject the null hypothesis and conclude that the means are indeed different. If the value is greater than the level of significance chosen beforehand, then you fail to reject the null hypothesis and conclude that the means are not different. In our example, the level of significance chosen was .05. Since the obtained significance value is .000, which is smaller than .05, we conclude that the difference between the means of the counseling and control groups is statistically significant. This means that the mean difference between the two groups is unlikely due to chance. You cannot claim that your results are significant at the .001 level, though; to do that, you'd have to say ahead of time that you would only reject H_0 if you got such an extreme value. You must decide what level of significance will justify rejecting H_0 before you look at your data; that's the rule. But you're still okay; you got what you were looking for, a *t* that will allow you to reject H_0 at the .05 level. You conclude that the differences between your samples of counseled and control subjects reflect a real difference in the populations of counseled and noncounseled people. Another way of saying this is that there are fewer than 5 chances out of 100 that a value of *t* this large could have occurred by chance alone. The statistician's way of saying it is $p < .05$.

The level of significance chosen is known as α (alpha). Why α, rather than *p*? It's a seemingly minor difference, but α is the probability of a Type I error, and *p* refers to the probability of getting the actual results you got just by chance. The most typical alpha level for social science research is $\alpha = .05$. As indicated in

Chapter 10, you might choose $\alpha = .01$ if you want to be super careful to avoid committing a Type I error. If a significant result would commit you to investing a great deal of money in program changes, or would lead to other important policy decisions, for example, then a Type I error would be quite dangerous and you would want to be very cautious indeed in setting your α level. Got it?

In addition to these results, you can also see the results of Levene's test for equality of variances. As mentioned above, for the *t* test results to be valid, we have to assume that the two samples come from populations with the same variances. Levene's test examines this assumption. As we can see here, the value of Levene's *F* statistic (we'll talk more about *F* tests in the Chapter 12) is 3.393 and has a significance value of .079. Since this value is greater than .05, we fail to reject the null hypothesis and assume that the variances of the populations are the same. If the variances of the two groups were different, then you would simply use the value of *t* that does not assume equal variances.

The last thing I want to point out is that the abbreviation *df* means "degrees of freedom." To find the degrees of freedom for a *t* test for independent samples, just subtract 2 from the total number of subjects in your study. In our example, $df = n_1 + n_2 - 2 = 12 + 12 - 2 = 22$. Before the use of statistical software programs like SPSS, the *df* was used along with *p* to find a critical value of *t* in a table of statistics. The critical value of *t* was then compared to the obtained value of *t* to determine whether the results of the test were statistically significant. Finding the critical value of *t* is no longer necessary, however, because we can directly examine the *p* value for any given test.

Let's do one more example, going through each step in a typical study comparing two groups. Suppose you wanted to test the hypothesis that men and women differ in the degree of empathy that they show to a stranger. You could select representative samples of men and women, tape-record their con-

**"In an increasingly complex world,
sometimes old questions require new answers."**

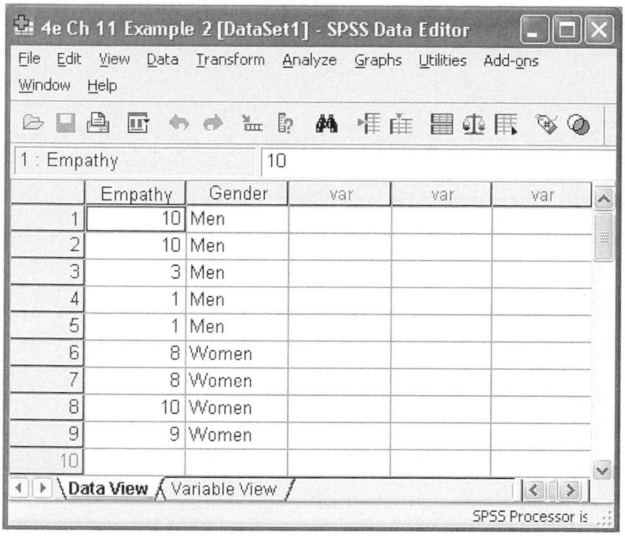

Fig 11-5 Example results.

versations with your research associate, and use some sort of test to measure their degree of empathy. Imagine that your results were as shown (the higher the score, the greater the degree of empathy) in Figure 11-5.

Step 1: State Your Hypothesis.   The statistical hypothesis you will be testing is the null hypothesis. In this example, the null hypothesis is that there is no difference between populations of men (Group 1) and women (Group 2) in the level of empathy that they offer to a stranger. In statistical terms,

$$H_0 : \mu_1 = \mu_2$$

Sometimes hypotheses are stated as alternative or research hypotheses, which represent the thing you want to show to be true. Alternative or research hypotheses are the opposite of the null hypothesis. In this case, the alternative hypothesis would be that there is a difference between populations of men and women in the degree of empathy that they offer to a stranger:

$$H_a : \mu_1 \neq \mu_2$$

Step 2: Select α, Your Significance Level.   Let's choose $\alpha = .05$.

Step 3: Compute t.   To conduct the analysis, click on *Analyze* in the menu bar; then on *Compare Means*; and finally on the *Independent-Samples T Test*. Now click on *Empathy* and then on the ▶ to move it to the *Variable(s)*: box. Next, click on *Gender* and then on the ▶ next to the *Grouping Variable*: area. After this, click *Define Groups* to bring up the *Define Groups* dialog box. Here, type in Men in the Group 1 box and Women in the Group 2 box. Now

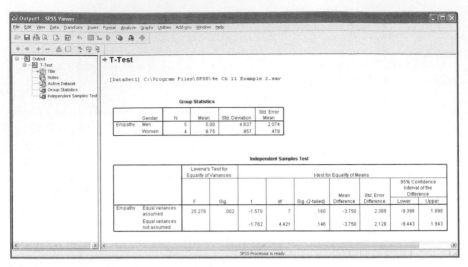

Fig 11-6 Independent-Samples *t* Test results.

click on *Continue* and then on *OK* in the *Independent-Samples T Test* dialog box to run the analysis. The results of this analysis are shown in Figure 11-6.

Step 4: Examine Levene's Test to Determine Appropriate t Test to Use.   As you can see in the figure, the *p* value of our test of the assumption of equal variances is .002. Since *p* < .05, we reject the hypothesis that the variances of the two samples are equal.

Step 5: Decide Whether or Not to Reject the Null Hypothesis. Because we cannot assume equal variances, we must use the results for the *t* test that does not assume equal variances in the bottom row of the figure. As I pointed out earlier, the sign of *t* will depend on which sample mean you happened to label $\bar{X}_1$ and which one you labeled X_2. For the hypothesis you are testing now, it doesn't really matter which sample mean is larger; you're only interested in whether or not they're different. If *p* < .05, you will reject H_0. As you can see in the figure, our *p* value is .146. Given that .146 > .05, we fail to reject the null hypothesis at the .05 level of significance. Despite the fact that the actual values of the means look quite different, based on these results it is *not* reasonable to conclude that female counselors offer a higher level of empathy to their clients than do male counselors.

THE *t* TEST FOR DEPENDENT (MATCHED) SAMPLES

Suppose you gave a group of ten participants a test both before and after a movie intended to influence attitudes toward public schools. You had two sets of scores, one from the pretest and the other from the posttest, and you wanted to find out if attitudes as measured by the tests were more or less favorable after seeing the movie than they were before. You now have pairs of scores, a score for each subject on the pretest, and another score from each subject for the posttest. You have two groups of scores, but they are not independent of each other; they are matched, as shown in Figure 11-7.

The most important thing to notice about these data, when deciding what statistical test to use, is that the scores are paired. The pretest score of 84 goes with the posttest score of 89, and it is the fact that Participant 1 raised his or her score by 5 points that is important, rather than the values of the two scores by themselves. It wouldn't make sense to scramble the posttest scores and then look at the difference between pretest and posttest scores. Each pretest score is logically linked to one, and only one, posttest score. That's the definition of dependent samples; whenever that condition holds, then a nonindependent samples test is appropriate.

Here is the formula for the dependent-samples *t* test:

$$t^*_{\text{obt}} = \frac{\bar{X}_1 - \bar{X}_2}{\sqrt{SD^2_{\bar{X}_1} + SD^2_{\bar{X}_2} - 2r_{12}SD_{\bar{X}_1}SD_{\bar{X}_2}}}$$

Where: $\bar{X}_1, \bar{X}_2$ are the means of the two measurements,

$SD_{\bar{X}_1}SD_{\bar{X}_2}$ are the standard errors of the means $(SD_X/\sqrt{N})$, and

r_{12} is the correlation between the two variables.

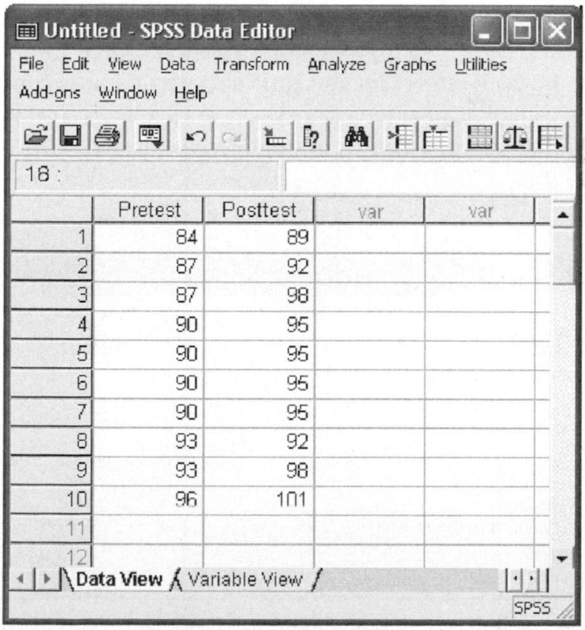

	Pretest	Posttest	var	var
1	84	89		
2	87	92		
3	87	98		
4	90	95		
5	90	95		
6	90	95		
7	90	95		
8	93	92		
9	93	98		
10	96	101		
11				
12				

Fig 11-7 Dependent-sample data.

Here we go, step by step, to conduct the dependent samples *t* test:

Step 1: State Your Hypothesis.   Your null hypothesis is that there is no difference between attitudes before the movie and attitudes after the movie; that is,

$$H_0 : \mu_1 = \mu_2$$

Your alternative, or research, hypothesis is that there is a difference between attitudes before and after the movie:

$$H_a : \mu_1 \neq \mu_2$$

Step 2: Select Your Level of Significance.   For this example, we will select the .01 level of significance, $\alpha = .01$.

Step 3: Compute t.   To conduct the *t* test for dependent samples, click on the *Analyze* menu in the menu bar; then click on the *Compare Means* item in the menu; and finally on the *Paired-Samples T Test*. Now click on *Pretest* and then on *Posttest* so that they are both highlighted. Notice that they are listed in the *Current Selections* area as Variable 1 and Variable 2 as shown in Figure 11-8.

Once you have done this, click on the ▶ to move it to the *Paired Variables*: box and then click *OK*. Results of this analysis are shown in Figure 11-9.

Step 4: Decide Whether to Reject H_0.   As was the case with independent *t* tests, we compare our obtained *p* value with our predetermined α level of .01. If our *p* value is less than .01, then we reject H_0 and conclude that our results are significant at the chosen level of α. Since in our example the obtained value is .000, we reject the null hypothesis and conclude that the movie appeared to have a positive effect on attitudes toward public schools. We don't conclude that attitudes were unchanged—that would be accepting H_0, and we can't do that.

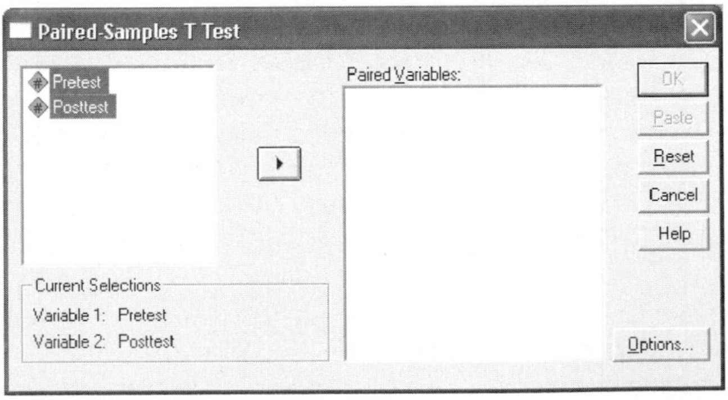

Fig 11-8 Paired samples *t* Test.

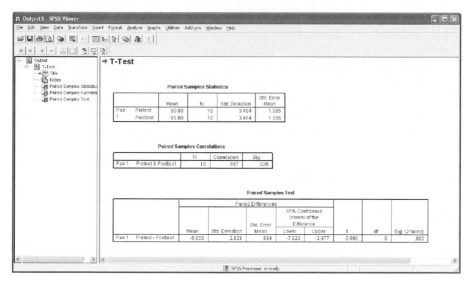

Fig 11-9 Paired samples *t* Test results.

DIRECTIONAL VERSUS NONDIRECTIONAL TESTS

When you as a researcher are quite confident, on the basis of previous research or theory, that the mean of one group should be higher than that of some other group, and you predict the direction of the difference before you collect your data, you can then use what is called a directional, or one-tailed, *t* Test. In a directional test, the alternative hypothesis would be stated like this if you expected the mean of Group 1 to be larger than Group 2:

$$H_\alpha : \mu_1 > \mu_2$$

And like this if you expected the mean of Group 1 to be smaller than Group 2:

$$H_\alpha : \mu_1 < \mu_2$$

The formula and output for directional (one-tailed) and non-directional (two-tailed) *t* Tests are exactly the same. However, as you may have noticed, SPSS only provides the p-value for a two-tailed *t* Test in the column labeled *Sig. (two-tailed)*. If you're conducting a one-tailed test, you must divide the obtained p-value by 2 before deciding to reject the null hypothesis or not. For example, if you conducted a one-tailed test with α = .05 and obtained a p-value of .06, you would divide it by 2 like this: .06 / 2 = .03. Since p = .03 is less than .05 you would reject the null hypothesis. As you can see, it's easier to reject the null hypothesis with a one-tailed test than it is with a two-tailed test.

So why not use a one-tailed test all the time? To use the one-tailed test legitimately, you must make your predictions prior to data collection. To do otherwise would be analogous to placing your bets after you see the outcome of an event. When in doubt, it is better to do two-tailed tests, if only to avoid temptation. However, doing a two-tailed test does increase the likelihood of a Type II error, that is, of not rejecting the null hypothesis when it should be rejected. If a significant outcome of your research would make sense only if the observed differences are in a particular direction (if you'd dismiss anything else as chance or random differences, no matter how unlikely), then do a one-tailed test. Remember, though, that if you choose to do a one-tailed test and your data show "significant" differences in the opposite direction than predicted, you may not reject H_0. By choosing a one-tailed approach, you have committed yourself to assuming that any differences in the non-predicted direction are due to chance or error.

PROBLEMS

For each problem, be sure to specify the null hypothesis being tested, and whether you will use a *t* test for independent samples or a *t* test for dependent samples.

1. In a study designed to discover whether men or women drink more coffee, a researcher (working on a very limited budget) observes five men and five women randomly selected from her university department. Here's what she found:

Run the appropriate test with $\alpha = .05$, assuming that both men and women were originally part of one random sample, with $n = 10$, and were then divided into men's (Group 1) and women's (Group 2) groups. What do you conclude based on your results?

2. Using hospital and agency records, you locate six pairs of identical twins, one of whom was adopted at birth and the other of whom was in foster care for at least 3 years. All the twins are now 5 years old. You are interested in knowing whether there is a difference in intelligence between the Adopted (Group 1) and Foster Care (Group 2) groups, so you test all the twins with the Wechsler Intelligence Scale for Children (WISC). Your results are as follows:

Run the appropriate test with $\alpha = .05$.

ANSWERS TO PROBLEMS

1. $H_0 : \mu_1 = \mu_2$; t test for independent samples; do not reject H_0. These results suggest that men and women drink the same amount of coffee.

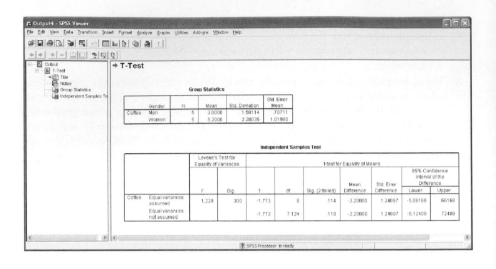

T-Test

Group Statistics

		Gender	N	Mean	Std. Deviation	Std. Error Mean
Coffee	Men		5	3.0000	1.58114	.70711
	Women		5	5.2000	2.28035	1.01980

Independent Samples Test

		Levene's Test for Equality of Variances		t-test for Equality of Means						
									95% Confidence Interval of the Difference	
		F	Sig.	t	df	Sig. (2-tailed)	Mean Difference	Std. Error Difference	Lower	Upper
Coffee	Equal variances assumed	1.228	.300	-1.773	8	.114	-2.20000	1.24097	-5.06168	.66168
	Equal variances not assumed			-1.773	7.124	.119	-2.20000	1.24097	-5.12408	.72408

2. $H_0: \mu_1 = \mu_2$; t test for dependent samples; reject H_0. These results suggest that the mean intelligence of twins in adopted families and foster families is not the same.

T-Test

Paired Samples Statistics

		Mean	N	Std. Deviation	Std. Error Mean
Pair 1	Adopt	106.83	6	9.663	3.945
	Foster	91.33	6	6.831	2.789

Paired Samples Correlations

		N	Correlation	Sig.
Pair 1	Adopt & Foster	6	-.526	.284

Paired Samples Test

		Paired Differences					t	df	Sig. (2-tailed)
					95% Confidence Interval of the Difference				
		Mean	Std. Deviation	Std. Error Mean	Lower	Upper			
Pair 1	Adopt - Foster	15.500	14.474	5.909	.310	30.690	2.623	5	.047

ENDNOTES

1. If we are to be very precise here, we would say that the difference between the means would be significant at the .05 level or at the .01 level.
2. Don't worry too much about what "two-tailed" means. In short, you can conduct one-tailed or two-tailed tests of statistical significance. The statistical tests themselves are identical, but the criterion used for determining significance differs for each kind of test. Because most researchers routinely conduct two-tailed tests of significance, we'll focus on two-tailed tests in this book. I'll say a little bit more about one and two tail tests at the end of the chapter, however.

12

Analysis of Variance (ANOVA)

- Analysis of Variance (ANOVA)
- Conducting an ANOVA in SPSS
- Strength of Association
- Post hoc Analyses
- The Scheffé Method of Post hoc Analysis
- Problems

The Top Ten Reasons Why Statisticians Are Misunderstood

1. They speak only the Greek language.
2. They usually have long, threatening names such as Bonferonni, Scheffé, Tchebycheff, Schatzoff, Hotelling, and Godambe. Where are the statisticians with names such as Smith, Brown, or Johnson?
3. They are fond of all snakes and typically own as a pet a large South American snake called an ANOCOVA.
4. For perverse reasons, rather than view a matrix right side up they prefer to invert it.
5. Rather than moonlighting by holding Amway parties, they earn a few extra bucks by holding pocket-protector parties.
6. They are frequently seen in their backyards on clear nights gazing through powerful amateur telescopes looking for distant star constellations called ANOVAs.
7. They are 99% confident that sleep cannot be induced in an introductory statistics class by lecturing on Z-scores.
8. Their idea of a scenic and exotic trip is traveling 3 standard deviations above the mean in a normal distribution.
9. They manifest many psychological disorders because as young statisticians many of their statistical hypotheses were rejected.
10. They express a deap-seated fear that society will someday construct tests that will enable everyone to make the same score. Without variation or individual differences the field of statistics has no real function and a statistician becomes a penniless ward of the state.

In Chapter 11, you learned how to determine if the means of two groups differ to a statistically significant degree. In this chapter, you will learn how to test for differences among the means of two *or more* groups. Hey, I bet you thought you were having fun before!

Suppose you assigned subjects to one of three groups: a peer support group (Group 1), an exercise/diet group (Group 2), and a no-treatment control group (Group 3), with post-treatment test scores on a measure of well-being, as shown in Figure 12-1.

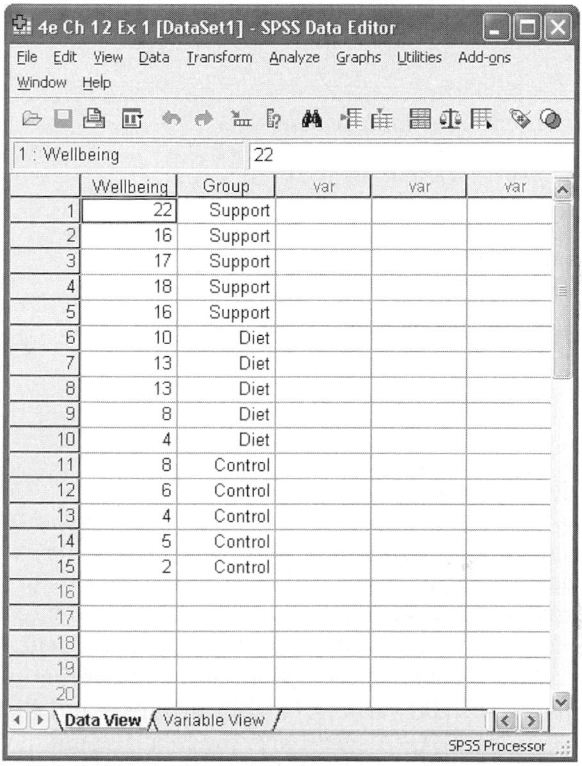

	Wellbeing	Group	var	var	var
1	22	Support			
2	16	Support			
3	17	Support			
4	18	Support			
5	16	Support			
6	10	Diet			
7	13	Diet			
8	13	Diet			
9	8	Diet			
10	4	Diet			
11	8	Control			
12	6	Control			
13	4	Control			
14	5	Control			
15	2	Control			
16					
17					
18					
19					
20					

Fig 12-1 Example ANOVA data.

You could test for the differences between pairs of means with the *t* Test: You could test for the significance of the difference for $\bar{X}_1$ versus $\bar{X}_2$, $\bar{X}_1$ versus $\bar{X}_3$, and $\bar{X}_2$ versus $\bar{X}_3$. There are at least two reasons why it would not be a good idea to do this kind of analysis, however.

1. It's tedious. If you do *t* Tests, you will have to compute $\frac{k(k-1)}{2}$ of them, where k is the number of groups. In our example, $\frac{k(k-1)}{2} =$

$\dfrac{3(3-1)}{2} = \dfrac{6}{2} = 3$. This isn't too bad, but if you were comparing means among, say, 10 groups, you would have to compute 45 t Tests!

2. More important, when you select, for example, the .05 level of significance, you expect to make a Type I error 5 times out of 100 by sampling error alone for each test. If you performed 20 t Tests and one of them reached the .05 level of significance, would that be a chance occurrence or not? What if 3 of the 20 were significant—which would be chance and which would be a "real" difference? With multiple t Tests, the probability (p) of a Type I error is

$$p = 1 - (1 - \alpha)^c$$

Where:　　α = alpha level
　　　　　c = number of comparisons

For our example, if $c = 3$ and $\alpha = .05$, then

$$p = 1 - (1 - \alpha)^c = 1 - (1 - .05)^3 = 1 - .95^3 = .14$$

So, instead of a 5% chance of making a Type I error, when you conduct three t Tests you have a 14% chance of committing a Type I error.

ANALYSIS OF VARIANCE (ANOVA)

You can take care of both of these problems by using *analysis of variance* (ANOVA) to test for statistical significance of the differences among the means of two or more groups. It may be important to note here that even though the name of this statistic has the term "variance" in it, it is used to test for significant differences among means. The test looks at the amount of variability (the differences) between the means of the groups, compared with the amount of variability among the individual scores in each group—that is, the variance between groups versus the variance within groups—and that's where the name comes from. The ANOVA starts with the total amount of variability (i.e., variance) in the data and divides it up (statisticians call it "partitioning") into various categories. Eventually, the technique allows us to compare the variability among the group means with the variability that occurred just by chance or error—and that's exactly what we need to be able to do.

Perhaps you recall the formula for the variance, given to you in Chapter 5. Remember that when we are estimating the variance of the population from which the sample was drawn, we must divide the sum of the deviation scores by $N - 1$, rather than just N: Just getting the average of the sample members' deviations around the mean would yield a biased estimate of the population value. $N - 1$, one less than the number of things in the sample (scores on a test, people, hot fudge sundaes) is known as the sample's *degrees of freedom*.

Degrees of freedom (df) is a concept you may not understand yet, although we've used the term several times already. The basic idea has to do with the number of scores in a group of scores that are free to vary. In a group of 10 scores that sum up to 100, you could let 9 of the scores be anything you wanted. Once you had decided what those 9 scores were, the value of the 10th score would be determined. Let's say we made the first 9 scores each equal to 2. They'd add up to a total of 18; if the sum has to be 100, then the 10th score has to be 82. The group of 10 scores has only 9 degrees of freedom, 9 scores that are free to vary, $df = 9$. Why is this important? Because the calculations for an ANOVA involve degrees of freedom, and you need to be able to figure out what those df are. But we need to do a few other things first.

The first step in carrying out an analysis of variance is to compute the variance of the total number of subjects in the study—we put them all together, regardless of the group to which they've been assigned, and find the variance of the whole thing. We do this using $N_T - 1$ (the total degrees of freedom) for the denominator of the formula:

$$SD_T^2 = \frac{\sum (X - \bar{X}_T)^2}{N_T - 1}$$

Nothing new so far—this is just our old friend, the formula for estimating a population variance based on a sample drawn from that population. We do have a couple of new names for things, though. The numerator of this formula is called the "total sum of squares," abbreviated SS_T—"total," because it's calculated across the total number of scores, combining all the groups. SS_T is the basis for all the partitioning that will follow. Notice, too, that the formula uses $\bar{X}_T$ as the symbol for the overall mean of all scores (some authors use "GM," for "grand mean"), and N_T, the total number of subjects. The denominator of the formula is known as the total degrees of freedom, or df_T. Translating the old variance formula to these new terms, we get

$$SD_T^2 = \frac{\sum (X - \bar{X}_T)^2}{N_T - 1} = \frac{SS_T}{df_T}$$

In ANOVA calculations, this pattern—dividing a sum of squares by an associated df—is repeated again and again. The number that you get when you divide a sum of squares by the appropriate df is called a mean square (MS). So,

$$SD_T^2 = MS_T = \frac{SS_T}{df_T}$$

I want to pause here to remind you of something I said way back at the very beginning of this book: Mathematical formulas take much longer to read and understand than do most other kinds of reading. You worked through few formulas in Chapter 11, and we're going to be dealing with a few more of them here. So remember to take your time! Pause, translate the formula into words, and make sure you understand how it relates to what went before. This last formula, for instance, says that the total mean square of a group of scores is the sum of squares for that group, divided by the degrees of freedom. And what are the sum of squares, and the degrees of freedom? Go back, read again, and put it together in your head. Understand each piece before you go on to the next. Reading in this way will actually prove to be a faster way to learn, in the long run.

In a simple ANOVA, the total sum of squares (SS_T) is broken down into two parts: (a) a *sum of squares within groups*, SS_W, which reflects the degree of variability within groups, but is not sensitive to overall differences between the groups; and (b) a *sum of squares between groups*, SS_B, which reflects differences between groups but is not sensitive to variability within groups.

The SS_W is found by:

$$SS_W = \sum (X - \bar{X}_i)^2$$

Where: $\bar{X}_i$ is the mean of a group.

This formula squares the difference between each score in a group and the mean of that group, and then the squared differences across all groups are summed.

The SS_B is calculated by:

$$SS_B = \sum n_i (\bar{X}_i - \bar{X}_T)^2$$

Where: n_i is the number of scores in a group.

This formula squares the difference between the mean of each group ($\bar{X}_i$) and the grand mean ($\bar{X}_T$) and then sums these squared differences. By multiplying each squared difference by n_i the formula essentially counts the difference for each participant in a group. (See the Summation section in Appendix A if this formula is confusing.)

The total sum of squares is the sum of the sum of squares within and the sum of squares between:

$$SS_T = SS_W + SS_B$$

The total degrees of freedom can be broken down as well:

$$df_T = df_W + df_B$$

To find df_W, add up the df's within all the groups:

$$df_W = (n_1 - 1) + (n_2 - 1) + \cdots + (n_{\text{Last}} - 1)$$

And df_B is the number of groups minus 1: $k - 1$.

For the groups in our example,

$$df_W = (5 - 1) + (5 - 1) + (5 - 1) = 4 + 4 + 4 = 12$$
$$df_B = 3 - 1 = 2$$

If we did our math right, $df_W + df_B$ should equal df_T:

$$12 + 2 = 14$$

Dividing SS_W by df_W gives us what is known as the mean square within, a measure of the variability within groups:

$$MS_W = \frac{SS_W}{df_W}$$

And dividing SS_B by df_B gives us the mean square between, a measure of variability between groups:

$$MS_B = \frac{SS_B}{df_B}$$

With MS_B, we have a measure of variability between the groups, that is, a measure that reflects how different they are from each other. And with MS_W we have a measure of the variability inside the groups, that is, variability that can be attributed to chance or error. Ultimately, of course, we want to know if the between-group differences are significantly greater than chance. So we will compare the two by computing their ratio:

$$F = \frac{MS_B}{MS_W}$$

The F statistic is the ratio of a mean square between groups to a mean square within groups. (It's named after Sir Ronald Fisher, who invented it.) As usual, we will compare the p value for the obtained value of F with a predetermined level of significance (e.g., $\alpha = .05$) to decide whether to reject H_0 or not. If $p < .05$, then we reject the null hypothesis.

You may have guessed that when comparing three group means the null hypothesis is

$$H_0 : \mu_1 = \mu_2 = \mu_3$$

Assumptions for Using ANOVA

The three assumptions underlying the use of one-way ANOVA are same as those for the t test for independent samples: independence, normality, and equality of variances. To meet the assumption of independence, the score for any particular participant must be independent from the score for any other participant. For the normality assumption to be met, the population from which the participants are sampled must be normally distributed. This assumption is robust to violations when an independent variable has a fixed

United Media/United Feature Syndicate, Inc. © 1979 PEANUTS reprinted by permission of United Feature Syndicate, Inc.

number of levels, however. Last, the equal variances assumption is also robust when cell sizes are equal. When cell sizes are not equal, there are tests that can be conducted to examine whether this assumption is met. We'll be using examples with equal cell sizes, however, so this won't be a problem for us.

CONDUCTING AN ANOVA IN SPSS

To conduct a one-way ANOVA, click on *Analyze* in the menu bar; then click on *General Linear Models*, and finally on *Univariate. . . .* This will bring up the *Univariate* dialog box shown in Figure 12-2.

Just as we've done a number of times before, click on the variable *Wellbeing* and then on the ▶ next to the *Dependent Variable*: box to move it to

Fig 12-2 Univariate dialog box.

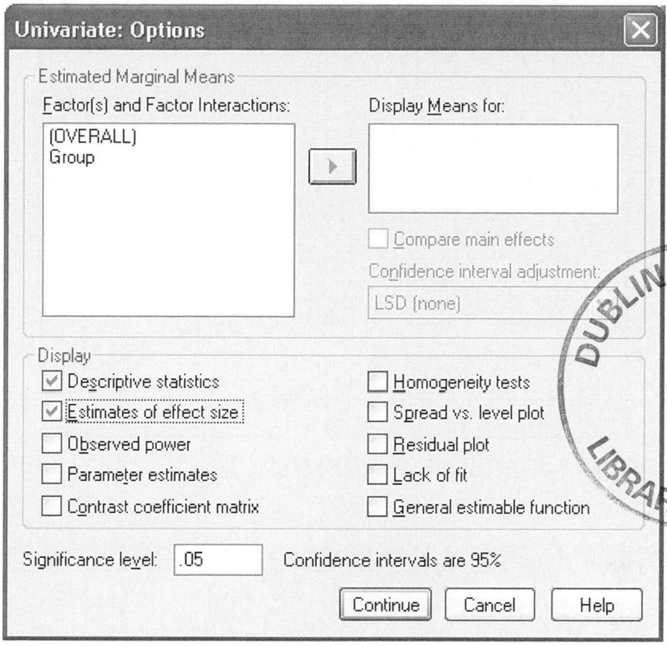

Fig 12-3 Univariate: Options dialog box.

that box. Now click on *Group* and then on the ▶ next to the *Fixed Factor(s)*: area. After you have done this, click on *Options* to bring up the *Univariate*: *Options* dialog box shown in Figure 12-3.

In this dialog box, click on *Group* in the *Factor(s) and Factor Interactions*: area and then on the ▶ to move the variable to the *Display Means for*: area. Also, select the boxes for both *Descriptive statistics* and *Estimates of effect size* and make sure you have .05 in the *Significant level* box. After you have done this, click *Continue* and then on *Post hoc . . .* in the *Univariate* dialog box. This will open the *Univariate*: *Post Hoc Multiple Comparisons for Observed Means* dialog box shown in Figure 12-4.

In this dialog box, click on *Group* in the *Factor(s)*: area and then on the ▶ to move the variable to the *Post Hoc Tests for*: area. In addition, select the box for *Scheffe* in the *Equal Variances Assumed* area and then click on *Continue*. Finally, in the *Univariate* dialog box, click on *OK* to run the analysis.

Figure 12-5 displays output for the ANOVA. The first two boxes show the number of cases per group, or factor, and the descriptive statistics for each group.

As was true for the *t* Test, to find out whether your *F* is statistically significant, you will need to compare the *p* value for the *F* statistic with the α level of .05 that we decided upon before beforehand. As you can see in the figure, $F = 24.729$ and the corresponding *p* value, in the column titled *Sig.*, is .000.

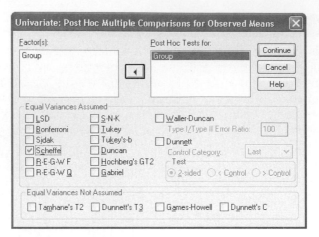

Fig 12-4 Univariate: Post hoc dialog box.

Univariate Analysis of Variance

[DataSet1] C:\Program Files\SPSS\4e Ch 12 Ex 1.sav

Between-Subjects Factors

		N
Group	Control	5
	Diet	5
	Support	5

Descriptive Statistics

Dependent Variable: Wellbeing

Group	Mean	Std. Deviation	N
Control	5.00	2.236	5
Diet	9.60	3.782	5
Support	17.80	2.490	5
Total	10.80	6.109	15

Tests of Between-Subjects Effects

Dependent Variable: Wellbeing

Source	Type III Sum of Squares	df	Mean Square	F	Sig.	Partial Eta Squared
Corrected Model	420.400[a]	2	210.200	24.729	.000	.805
Intercept	1749.600	1	1749.600	205.835	.000	.945
Group	420.400	2	210.200	24.729	.000	.805
Error	102.000	12	8.500			
Total	2272.000	15				
Corrected Total	522.400	14				

a. R Squared = .805 (Adjusted R Squared = .772)

Fig 12-5 ANOVA results.

Because the $p < .05$, we conclude that the result is statistically significant and reject H_0.

A statistically significant F test in ANOVA tells us that *at least* one of the means is significantly different from one of the others, but it doesn't tell us which means are significantly different. Eyeballing the means of the three groups in our example, we can see that the mean of Group 1, the peer support group ($\bar{X}_1 = 17.80$), is much larger than the mean of Group 3, the control group ($\bar{X}_2 = 5.0$). It seems likely that these two means are statistically significantly different. We also might wonder, however, if the mean of Group 1 is different from the mean of Group 2, the exercise/diet group ($\bar{X}_3 = 9.6$), and if the exercise/diet group is significantly different from the control group. To answer these questions, you will need to learn how to do what is known as *post hoc analysis*. But first, let's talk about strength of association.

STRENGTH OF ASSOCIATION

With ANOVA, the overall F test tells us whether the difference between at least one of the means of our groups was statistically significantly different. The F test does not, however, tell us anything about the strength of the treatment effect. With very large samples, you often find differences between means that are statistically significant, but not very important in terms of practical significance. For this reason, whenever you find a statistically significant F test, you also need to examine the strength of association for the treatment effects. In SPSS, you can find the proportion of total variability accounted for by the treatment effects with *eta-squared*. When we selected the *Estimates of Effect Size* in the *Univariate*: Options dialog box (see Figure 12-3), we told SPSS to provide us with this statistic. Selecting this provides a partial eta-squared value for each factor, which in our case refers to the three treatment groups. The partial eta-squared is interpreted in the same way as the coefficient of determination, r_{xy}^2. If that concept is a bit fuzzy, go back to Chapter 8 for a quick review. For our example, the partial eta-squared is 0.805. This indicates that the treatment effects in our example account for a large amount of variance—about 80% in fact—in post-treatment adjustment test scores.

POST HOC ANALYSES

Post hoc analyses are used after an ANOVA has been done and the null hypothesis of no difference among means has been rejected. Let's look at another example. Consider a study of five different teaching methods. Five groups of students were taught a unit, each group being exposed to a different teaching method, and then the groups of students were tested for how much they had learned. Even if an ANOVA were to show the differences among the five means to be statistically significant, we still would not know which of the pairs of means were significantly different: Is $\bar{X}_1$ significantly different from $\bar{X}_2$? How about the difference between $\bar{X}_2$ and $\bar{X}_3$? If we looked at each possible combination, we would have $5(5 - 1)/2 = 10$ pairs of means to analyze. You will recall from the discussion at the beginning of this chapter that it is not good practice to analyze differences among pairs of means with the t test because of the increased probability of a Type I error; the same criticism can be leveled at any large set of independent comparisons.

Many procedures have been developed to do what is called post hoc analysis (tests used after an F test has been found to be statistically significant). This book will present only one of these methods, the Scheffé method, which can be used for groups of equal or unequal N's.

THE SCHEFFÉ METHOD OF POST HOC ANALYSIS

The statistic you will compute in the Scheffé method is designated as C. A value of C is computed for any contrast, or pair of means that you want to compare; unlike the t test, C is designed to allow multiple comparisons without affecting the likelihood of a Type I error. Moreover, if all the groups are the same size, you don't have to compute C for every single pair of means; once a significant C has been found for a given pair, you can assume that any other pair that is at least this far apart will also be significantly different. (With a t test, this is not necessarily true; nor is it always true for C when the group N's are unequal.)

As is the usual procedure, you will compare your C with a predetermined level of significance, such as $\alpha = .05$. If $p < .05$, then you reject the null hypothesis for that pair of means. Here is the formula for the Scheffé test:

$$C = \frac{\bar{X}_1 - \bar{X}_2}{\sqrt{MS_W\left(\dfrac{1}{n_1} + \dfrac{1}{n_2}\right)}}$$

Where: $\bar{X}_1, \bar{X}_2$ are the means of two groups being compared
n_1, n_2 are the n's of those two groups
MS_W is the within-group mean square from your ANOVA

Now let's go back to our original three groups: peer support (Group 1), exercise/diet (Group 2), and control (Group 3). Since we were able to reject H_0, we know that at least one group is significantly different from one other group; but we don't know which groups they are. And there may be more than one significant difference; we need to check that out. In our example, as you

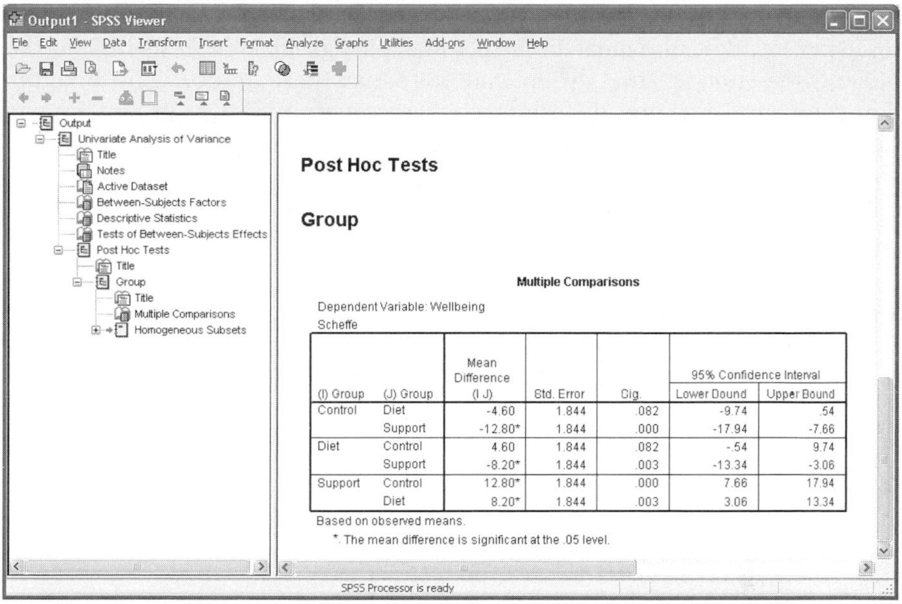

Multiple Comparisons

Dependent Variable: Wellbeing
Scheffe

(I) Group	(J) Group	Mean Difference (I-J)	Std. Error	Sig.	95% Confidence Interval Lower Bound	Upper Bound
Control	Diet	-4.60	1.844	.082	-9.74	.54
	Support	-12.80*	1.844	.000	-17.94	-7.66
Diet	Control	4.60	1.844	.082	-.54	9.74
	Support	-8.20*	1.844	.003	-13.34	-3.06
Support	Control	12.80*	1.844	.000	7.66	17.94
	Diet	8.20*	1.844	.003	3.06	13.34

Based on observed means.
*. The mean difference is significant at the .05 level.

Fig 12-6 Results of Scheffé post hoc analysis.

may remember, we selected the *Scheffe* option in the *Univariate: Post Hoc Multiple Comparisons for Observed Means* dialog box (see Figure 12-4). The output for this analysis is shown in Figure 12-6.

As you can see in the figure, the output for this analysis consists of multiple contrasts—each group is compared with the other two groups. The table displays the mean difference between groups and the statistical significance of each contrast.

Since it doesn't matter which group mean is subtracted from which in each computation of C, the sign of the value you get doesn't matter either. Just treat C as if it were positive. As you can see by examining results with the .05 level of significance, the mean of the support group is significantly larger than both the exercise/diet and the control groups. However, the mean of the exercise/diet group is not significantly greater than that of the control group.

Once you have determined that two groups are significantly different from each other, you will still have to decide if the differences are large enough to be useful in the real world. This is the "magnitude of effect" decision, which we discussed in relation to the t test. Unfortunately, there's no handy-dandy rule or formula to tell us whether or not we have a "large enough" magnitude of effect. It depends upon what's at stake: what's to be gained by a correct decision and what's to be lost by an incorrect one, and how much difference between groups is enough to be worth paying attention to. In our example, the mean of the peer support group is twice as high as the mean of the exercise/diet group, and more than three times the mean of the control group. For me, differences of that magnitude would play a very important part in treatment recommendations to students or clients. How about for you?

PROBLEMS

1. A researcher is interested in differences among blondes, brunettes, and redheads in terms of introversion/extroversion. She selects random samples from a college campus, gives each participant a test of social introversion, and comes up with the following:

Use a one-way ANOVA with $\alpha = .01$ to test for differences among the groups. What does she conclude, and how does she explain it?

2. Farmer MacDonald suspects that his chickens like music, because they seem to lay more eggs on days when his children practice their band instruments in the hen house. He decides to put it to a scientific test and records the number of eggs collected each day for a month, along with the music provided on that day:

```
┌──────────────────────────────────────────────────────────────┐
│ 🔲 4e Ch 12 Prob 2 [DataSet1] - SPSS Data Editor    [_][□][X]  │
│ File  Edit  View  Data  Transform  Analyze  Graphs  Utilities  Add-ons │
│ Window  Help                                                   │
│ ▷ 🖫 🖨 🖽 ↶ ↷ ⚏ ᦖ 🏥 ⫧ ⫨ 🔳 ⚖ ⧘ ◈ ◉                         │
│ 1 : Eggs                    │11                                │
```

	Eggs	Music	var	var	var
1	11	NoMusic			
2	26	NoMusic			
3	31	NoMusic			
4	18	NoMusic			
5	12	U2			
6	17	U2			
7	19	U2			
8	25	U2			
9	35	Gorillaz			
10	42	Gorillaz			
11	31	Gorillaz			
12	33	Gorillaz			
13	52	Aqualung			
14	78	Aqualung			
15	41	Aqualung			
16	73	Aqualung			
17					
18					
19					

```
◀ ▶ \Data View ⟨ Variable View /              ◁ ▷
                                       SPSS Processor
```

Use a one-way ANOVA with $\alpha = .05$ to test for differences among the groups. What does he conclude, and how does he explain it?

ANSWERS TO PROBLEMS

1. Since the $p > .01$ we cannot reject the H_0. Therefore, we know that none of the means is significantly different from another. There is no need for post hoc comparisons.

Univariate Analysis of Variance

[DataSet1] C:\Program Files\SPSS\4e Ch 12 Prob 1.sav

Between-Subjects Factors

		N
Group	Blonde	5
	Brunette	5
	Redhead	5

Tests of Between-Subjects Effects

Dependent Variable: Int_Ext

Source	Type III Sum of Squares	df	Mean Square	F	Sig.
Corrected Model	21.733[a]	2	10.867	2.131	.161
Intercept	232.067	1	232.067	45.503	.000
Group	21.733	2	10.867	2.131	.161
Error	61.200	12	5.100		
Total	315.000	15			
Corrected Total	82.933	14			

a. R Squared = .262 (Adjusted R Squared = .139)

2. Results of the one-way ANOVA are significant at the .05 level, indicating that at least two of the group means differ significantly. In addition, the partial eta-squared is large. The group explains 78% of the variance in the number of eggs laid. Results of *Scheffé* post hoc analysis indicate that the chickens laid significantly more eggs while listening to Aqualung than listening to U2, Gorillaz, and no music. No differences were found among the latter three conditions.

- Output
  - Univariate Analysis of Variance
    - Title
    - Notes
    - Active Dataset
    - Between-Subjects Factors
    - Descriptive Statistics
    - Tests of Between-Subjects Effects
    - Post Hoc Tests

Univariate Analysis of Variance

[DataSet1] C:\Program Files\SPSS\4e Ch 12 Prob 2.sav

Between-Subjects Factors

		N
Music	Aqualung	4
	Gorillaz	4
	NoMusic	4
	U2	4

Descriptive Statistics

Dependent Variable: Eggs

Music	Mean	Std. Deviation	N
Aqualung	61.00	17.455	4
Gorillaz	35.25	4.787	4
NoMusic	21.50	8.813	4
U2	18.25	5.377	4
Total	34.00	19.735	16

Tests of Between-Subjects Effects

Dependent Variable: Eggs

Source	Type III Sum of Squares	df	Mean Square	F	Sig.	Partial Eta Squared
Corrected Model	4539.500[a]	3	1513.167	13.941	.000	.777
Intercept	18496.000	1	18496.000	170.405	.000	.934
Music	4539.500	3	1513.167	13.941	.000	.777
Error	1302.500	12	108.542			
Total	24338.000	16				
Corrected Total	5842.000	15				

a. R Squared = .777 (Adjusted R Squared = .721)

- Output
  - Univariate Analysis of Variance
    - Title
    - Notes
    - Active Dataset
    - Between-Subjects Factors
    - Descriptive Statistics
    - Tests of Between-Subjects Effects
    - Post Hoc Tests
      - Title
      - Music
        - Title
        - Multiple Comparisons
        - Homogeneous Subsets
          - Title
          - Eggs

Post Hoc Tests

Music

Multiple Comparisons

Dependent Variable: Eggs
Scheffe

(I) Music	(J) Music	Mean Difference (I-J)	Std. Error	Sig.	95% Confidence Interval	
					Lower Bound	Upper Bound
Aqualung	Gorillaz	25.75*	7.367	.033	1.91	49.59
	NoMusic	39.50*	7.367	.002	15.66	63.34
	U2	42.75*	7.367	.001	18.91	66.59
Gorillaz	Aqualung	-25.75*	7.367	.033	-49.59	-1.91
	NoMusic	13.75	7.367	.365	-10.09	37.59
	U2	17.00	7.367	.205	-6.84	40.84
NoMusic	Aqualung	-39.50*	7.367	.002	-63.34	-15.66
	Gorillaz	-13.75	7.367	.365	-37.59	10.09
	U2	3.25	7.367	.977	-20.59	27.09
U2	Aqualung	-42.75*	7.367	.001	-66.59	-18.91
	Gorillaz	-17.00	7.367	.205	-40.84	6.84
	NoMusic	-3.25	7.367	.977	-27.09	20.59

Based on observed means.
*. The mean difference is significant at the .05 level.

13

Nonparametric Statistics: Chi-square

- Nonparametric Statistical Tests
- One-way Chi-square Test
- Conducting Chi-square in SPSS
- A Bit More
- Problem

Statisticians do it continuously but discretely.
Statisticians do it when it counts.
Statisticians do it with 95% confidence.
Statisticians do it with large numbers.
Statisticians do it with only a 5% chance of being rejected.
Statisticians do it with two-tail t tests.
Statisticians do it. After all, it's only normal.
Statisticians probably do it.
Statisticians do it with significance.
Statisticians do all the standard deviations.

"Nonparametric"—Oh, no! Is this something I should recognize from what has gone before? If we're going to look at nonparametric things, what in the world are parametric ones? Relax—all these questions will be answered. You should know, though, first of all, that the things we will be doing in this chapter will be easier, not harder, than what you have already done.

NONPARAMETRIC STATISTICAL TESTS

All the statistical procedures that you have learned so far are parametric procedures. The idea of parametric statistics involves a number of things, including the assumption that the data we work with are drawn from normally distributed populations, but the most important thing is that the data used in parametric tests or techniques must be scores or measurements of some sort. When comparing groups of people by means of a parametric test, we measure all the people and then use the test to determine whether the groups' measurements are significantly different from what could be expected just by chance.

But what if we are dealing with a situation in which we don't measure or test our subjects? That's not as unusual as you may imagine. Here's an example: A researcher wants to know if college students differ in terms of their use of the college counseling center by year in college. She randomly selects 100 students and asks them if they have ever used the center. Here are her findings, summarized by year in college:

Center Use	Frosh	Sophomore	Junior	Senior	Total
Used Center	20	31	28	21	100

How should this researcher analyze her data? There are no scores, no means, and no variances to calculate. What she has is four categories: students who used the center. Each person in the sample can be assigned to one, and only one, of these categories. What the researcher wants to know is whether the distribution she observed is significantly different from what she might expect, by chance, if the total population of students didn't differ by year.

http://www.borg.com/~rjgtoons/edu.html

THE ONE-WAY CHI-SQUARE (χ^2) TEST

Her question is quite easy to answer using a procedure called the chi-square test. Chi-square (symbolized by the Greek letter chi, squared: χ^2) is a nonparametric test. It doesn't require that its data meet the assumptions of parametric statistics, and it most particularly doesn't require that the data be in the form of scores or measurements. Instead, it was specifically designed to test hypotheses about categorical data. The one-way test of χ^2 can be used when the categories involve a single independent variable.

The null hypothesis for our example is that use of the counseling center doesn't differ significantly by year:

H_0 : The observed frequencies (f_0) equals the expected frequencies (f_e)

When H_0 is true, the differences between the observed and expected frequencies will be small. When H_0 is false, these differences will be relatively large. To test the H_0 using χ^2, we first need to figure out the distribution that we would most often get just by chance if H_0 were true. The expected frequencies for this study are those that would occur by chance if no difference in the use of the center existed across year in school. According to H_0, then, because there are 100 subjects, and we expect the same amount of students by year in school, we would predict 25 students to be freshmen, 25 sophomores, 25 juniors, and 25 seniors.

Center Use	Frosh	Sophomore	Junior	Senior	Total
Observed f	20	31	28	21	100
Expected f	25	25	25	25	100

Of course, in any experiment we would expect the observed frequencies to be slightly different from the expected frequencies on the basis of chance, or random sampling variation, even when H_0 is true. But how much variation from the expected frequencies is reasonable to expect by chance? When do we start to think that use of the center really does differ by year in school? Here's where the χ^2 can help.

The χ^2 provides a test of the discrepancy between expected and obtained frequencies:

$$\chi^2 = \Sigma \left[\frac{(O_i - E_i)^2}{E_i} \right]$$

Where: O_i is the observed frequency, and
E_i is the expected frequency

The χ^2 has degrees of freedom (df):

$$df = K - 1$$

Where: K is number of categories

By looking at the formula you can see a few things. First, the χ^2 will never be negative because we are squaring all differences between the expected and observed frequencies. Second, the χ^2 will equal zero only when the observed frequencies are exactly the same as the predicted frequencies. And third, the larger the discrepancies, the larger the χ^2. Let's do an example in SPSS, using a significance level of $\alpha = .05$.

CONDUCTING CHI-SQUARE IN SPSS

Before we go further, we need to learn one more thing about basic data entry and definition. Up to this point, we learned all we needed to cover about inputting and defining data in Chapter 3. So far, when we have had different groups in our examples we have simply labeled them by their names, such as male–female, treatment–control, and so on. However, we can also assign each group a number, such as: Male $= 0$ and Female $= 1$. Some researchers think it makes more sense to use numbers rather than words in data files. The easiest way to assign numbers to string variables in SPSS is to use the *Automatic Recode* procedure. From the menu bar, click on the *Transform* menu and then click on *Automatic Recode . . .* menu item to bring up the *Automatic Recode* dialog box shown in Figure 13-1.

Now click on *Year* and then on the ▶ to move the variable to the *Variable- > New Name* area. Next, type YearN in the *New Name* area to provide a name for

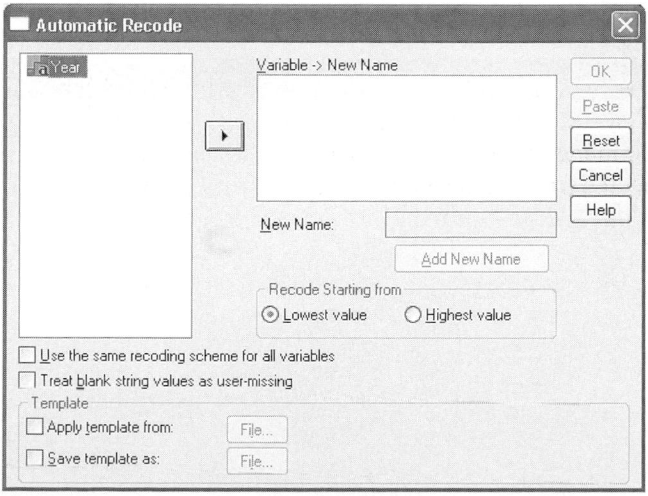

Fig 13-1 Automatic recode dialog box.

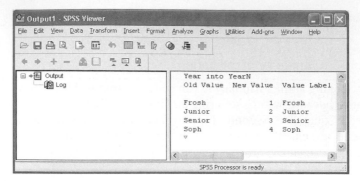

Fig 13-2 Automatic recode output.

the recoded variables, click on *Add New Name* button, and then click *OK*. Figure 13-2 displays the values for the new recoded variables.

The recoded variable is also listed as a new variable in the Data Editor with its respective values. Now we are ready to conduct our test.

ONE-WAY TEST OF CHI-SQUARE

To conduct a one-sample χ^2 in SPSS, click on the *Analyze* menu in the menu bar; then click on the *Nonparametric Tests* item in the menu; and finally on the *Chi-Square*. This will bring up the *Chi-Square Test* dialog box shown in Figure 13-3.

In the dialog box, click on *YearN*, our newly recoded variable, and then on the ▶ to move the variable to the *Test Variable List:* area. After this, click *OK* to run the procedure. Figure 13-4 displays the results of the χ^2 test.

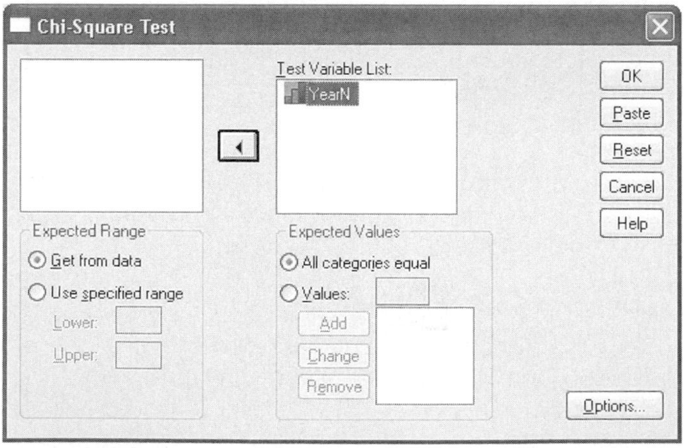

Fig 13-3 Chi-Square test dialog box.

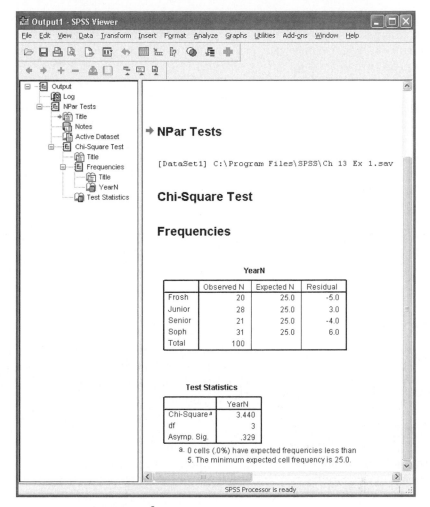

Fig 13-4 Results of the χ^2 test.

As shown in the figure, our observed value of χ^2 is 3.440 and the p value is .329. Since $p > .05$, we do not reject the null hypothesis and conclude that use of the counseling center does not differ by year in school.

It is important to note then when the test of χ^2 is statistically significant, this outcome would not imply any sort of causal relationship: We cannot say that year in school "causes" the differences in counseling center behavior, any more than we could say that use of the counseling center "causes" people to be either freshmen, sophomores, juniors, or seniors. Also, statistical significance does not guarantee magnitude of effect. Evaluation of magnitude of effect involves qualitative, as well as quantitative, considerations. Aren't you tired of having me say this?

Generally, studies utilizing χ^2 require a relatively large number of subjects. Many statisticians recommend a certain minimum expected frequency (f_e) per cell. A very conservative rule of thumb is that f_e must always be equal to or greater than 5.

A BIT MORE

This chapter has only focused on the one-way test of χ^2. There are lots more nonparametric tests out there—this is just a teaser to whet your appetite! For example, you can also use χ^2 with two or more levels. For examine, our researcher could examine whether men and women differed in their use of the counseling center by year in school. In general, it's safe to say that a nonparametric test exists, or can be cobbled to fit, virtually any research situation. Additionally, nonparametric tests tend to have a relatively simple structure and to be based on less complex sets of principles and assumptions than their parametric brethren. For this reason, if you can't find a ready-made nonparametric test that fits your situation, a competent statistician can probably design one for you without much trouble.

At this point, you may be asking, "So why don't we always use nonparametrics? Why bother with parametric tests at all?" The answer is simple: Parametric tests are more powerful than nonparametric statistics. When the data allow, we prefer to use parametrics because they are less likely to invite a Type II error. Another way of saying this is that parametric tests let us claim significance more easily; they are less likely to miss significant relationships when such relationships are present. However, using a parametric test with data that don't meet parametric assumptions can cause even more problems—can result in Type I error—and we can't even estimate the likelihood of a Type I error under these circumstances. That's when we need the nonparametric techniques. Not only are they, by and large, easier to compute, but they fill a much needed spot in our statistical repertoire.

PROBLEM

1. Senior education majors were asked about their plans for a job. Those who said they would like to teach the subject are represented below:

Teach in Junior College	Teach in a Teaching College	Teach in a Research University
22	19	9

What conclusion do you form based on the results? Are senior education majors' plans for a job related to higher-education institution? Test this hypothesis using $\alpha = .05$.

ANSWER TO PROBLEM

1. Do not reject the null hypothesis. Senior education majors' plans are not related to higher education institution.

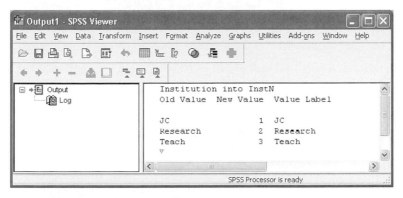

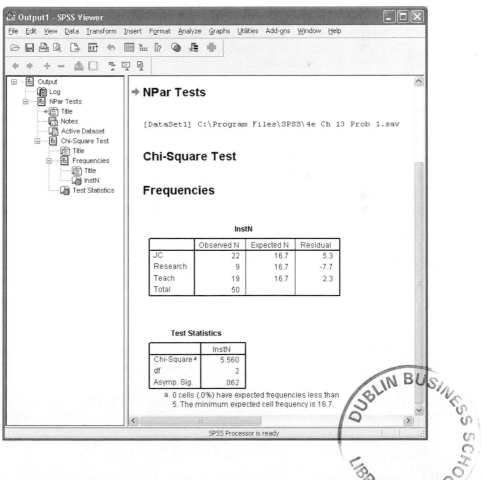

14

Postscript

- Congratulations!
- Review of Book

> Clem asks Abner, "Ain't statistics wonderful?" "How so?" says Abner. "Well, according to statistics, there's 42 million alligator eggs laid every year. Of those only about half get hatched. Of those that hatch, three-fourths of them get eaten by predators in the first 36 days. And of the rest, only 5 percent get to be a year old because of one thing or another. Ain't statistics wonderful?" Abner asks, "What's so wonderful about statistics?" "Why, if it wasn't for statistics, we'd be up to our asses in baby alligators!"

CONGRATULATIONS!

Well, you did it! You got all the way through this book! Whether you realize it or not, this means that you have covered a great deal of material, learned (probably) quite a bit more than you think you did, and are now able to do—or at least understand—most of the statistics that you will need for a large percentage of the research that you may become involved with. No small accomplishment!

In case you are inclined to discount what you've done, let's review it: This will not only give you further ammunition for self-congratulation but will also help to consolidate all the information you've been taking in.

REVIEW OF BOOK

Statistics, as a field of study, can be divided into two (not so equal) parts, descriptive and inferential. You've been introduced to both. First, the descriptive.

You've learned how to describe sets of data in terms of graphs (histograms, frequency polygons, cumulative frequency polygons), of central tendency (mean, median, mode), and of variability (range, variance, standard deviation).

You've learned that a distribution—and even that word was probably somewhat unfamiliar when you began all this—can be symmetrical or skewed, and you've learned what happens to the measures of central tendency when you skew a distribution. You also learned a lot about the properties of the normal curve and how to find proportions of scores in different areas of it. And you've learned how to use and interpret Z scores, T scores, and percentiles, as well as other standard scores.

You've also learned a lot about how two distributions—two sets of data—can be related. You learned how to compute a correlation coefficient and what a correlation coefficient means. You learned what a scatterplot is and how the general shape of a scatterplot relates to the value of r_{xy}. And you also learned how to compute a correlation coefficient on data comprised of ranks instead of measurements. You learned how to use a regression equation to predict a score on one variable, based on an individual's performance on a related variable, and to use something called the standard error of the estimate to tell you how much error to expect when making that prediction. That's a lot.

And then along came the inferential statistics: using a set of observable information to make inferences about larger groups that can't be observed. You started this section by absorbing a lot of general ideas and principles. You learned about probability, and how it relates to sampling. You learned why it's important that a sample be unbiased, and how to use random sampling techniques to get an unbiased sample. You learned what a null hypothesis is, why we need to use a null hypothesis, and what kinds of error are associated with mistakenly rejecting or failing to reject the null hypothesis. You learned that it's a great big no-no to talk about "accepting" the null hypothesis, and why! You learned what it means for a result to be statistically significant, and you got acquainted with a friendly Greek named α, as well as something called "magnitude of effect."

Then you moved into actual inferential statistical tests themselves, starting with the t test. Using the t test, you now know how to decide whether two groups are significantly different from each other, and you know that correlated or matched groups have to be treated differently from independent groups.

As if looking at two groups weren't enough, you moved right in to exploring comparisons among two or more groups. You learned about one of the workhorses of social science statistics, analysis of variance (ANOVA). And you learned how to do a post hoc test, to look at the means of those two or more groups in even more detail.

Finally, you learned a fine, important-sounding new word: *nonparametric*. You learned that nonparametric tests let you examine frequencies and ranked data. You learned how to use and interpret the most widely used nonparametric statistic—called chi-square. You now can work with data that don't fit the rules for the other techniques that you've learned.

You really have done a lot!

And we haven't even mentioned the single most important, and impressive, thing you've done. Imagine your reaction just a few months ago if someone had handed you the first paragraphs of this chapter and said, "Read this—this is what you will know at the end of this term." The very fact of your being able to think about statistics now without feeling frightened or overwhelmed or nauseous is much more significant than the facts and techniques that you've learned. Because your changed attitude means that you're able to actually use all this stuff, rather than just being intimidated by it. If you can't remember some statistical something now, you can go look it up, instead of giving up. If you can't find out where to look it up, you can ask somebody about it and have a reasonable expectation of understanding their answer.

Moreover, you're ready to move on to the next level: You've established for yourself a good, solid foundation that you can build on, just about as high as you want to go. There is more to statistics than we've been able to cover, of course. There are all the mathematical implications and "pre-plications" (well, what else do you call something that comes before and influences the thing you're interested in?) of the techniques you've learned. There are the fascinating nooks and crannies of those techniques—the sophisticated rules about when to use them, the exceptions to the rules, the suggestions for what to do instead. There are the extensions: applying the basic principles of correlation or regression or ANOVA to larger and more complex designs. And then there are the brand-new techniques, things like analysis of covariance, factor analysis, and multiple regression, among many others, and lots and lots of clever nonparametric tricks. Why, you might even learn to enjoy this stuff!

But whether you learn to enjoy it or not, whether you go on to more advanced work or just stay with what you now know, whether you actively use your statistics or simply become an informed consumer—whatever you do with it—nobody can change or take away the fact that you did learn it, and you did survive. And it wasn't as bad as you thought it would be—truly now, was it?

Appendix A

··

Basic Math Review

- Positive and Negative Numbers
- Fractions
- Decimals and Percents
- Exponents and Roots
- Order of Computation
- Summation

The appendix you're starting now will give you a chance to review the rules that you need in order to play with numbers and come up with the same answers as everyone else. Some of the material will be very familiar to you; other parts may seem completely new. Let me make a few suggestions about how to use this appendix:

1. If, after a couple of pages or so, you're completely bored and have found nothing you don't already know, just skim through the rest of A and get on with book.

2. If the material seems familiar, but you still feel a little shaky, that will tell you which parts you need to spend more time on.

3. If a lot of it feels new to you—take your time with it! Most of us "word people" absorb numerical information and ideas quite slowly and need lots of practice before it really sinks in and becomes part of our way of thinking. Give it a chance. Learning the rules now will allow you to understand the rest of the book in a way that will probably surprise you.

Learning to read statistics material is somewhat analogous to learning to read music or a foreign language: impossible at first, difficult for a while, but relatively easy after some effort. One thing to remember, though: Since symbols are a way of condensing information, a paragraph that is full of mathematical symbols has much more information in it than an ordinary paragraph in a history book or a newspaper article. Don't be surprised if it takes you three or four times longer to get through a page in a statistics book (even this one!)

than to get through a page in a nonnumerical book. In fact, one of the challenges for the beginning statistics student is learning to slow down. Force yourself to adjust your reading speed to the density of the information on the page, and you'll find that things get much easier.

I've divided the basic math rules into six sections: (a) positive and negative numbers, (b) fractions and percents, (c) roots and exponents, (d) order of computation, (e) summation, and (f) equations. Each section presents a number of important rules that should be followed when using statistics.

POSITIVE AND NEGATIVE NUMBERS

In a perfect and orderly world, all numbers would be positive (they'd be whole, too—no fractions or decimals). But the world isn't perfect, and negative numbers have to be dealt with. Actually, they're not so bad; you just have to show them who's boss. If you're a visually oriented person, it may help to think of numbers as standing for locations on a straight line, with zero as your starting point. The number 2 is two steps to the right from zero; add four more steps and you're six steps out, and so on. The negative numbers are just steps in the opposite direction. If I'm six positive steps (6 to the right) away from zero, and I subtract 4 from that, I take four steps back toward zero; now I'm at 2. If you're not visually oriented, the last paragraph may confuse you; if so, just ignore it and follow the rules I'm about to give you.

Rule 1

If a number or an expression is written without a sign, it's positive.

$$+2 = 2 \qquad +x = x$$

Rule 2

When adding numbers of the same sign, add them up and prefix them with the same sign as the individual numbers had.

$$+3 + (+5) = +8 \qquad 2 + 7 = 9$$

Rule 3

When summing up a group of numbers with mixed signs, think of the process as having three steps:

1. Add the positive numbers; add the negative numbers as if they were positive.
2. Subtract the smaller sum from the larger sum.
3. Prefix your answer with the sign of the larger sum.

$$5 - 3 + 2 - 1 \rightarrow (5 + 2) \text{ and } (3 + 1) = (7) \text{ and } (4)$$
$$7 \text{ (larger)} - 4 \text{ (smaller)} = 3$$

The answer is +3 because the positive sum (7) was larger than the negative sum. Here's another example:

$$-2 + 6 - 14 + 3 - 4 \rightarrow (6 + 3) \text{ and } (2 + 14 + 4) = (9) \text{ and } (20)$$
$$20 \text{ (larger)} - 9 \text{ (smaller)} = -11$$

The answer is –11 because the negative sum (20) was larger than the positive sum.

Rule 4

Subtracting a positive number is the same as adding a negative number; adding a negative number is the same as subtracting a positive number. Subtracting a negative is the same as adding a positive. In other words, two negative signs make a positive sign; a positive and a negative make a negative (you visually oriented thinkers, work it out on the number line).

$$5 - (+3) = 5 + (-3) = 5 - 3 = 2$$
$$7 + (-12) = 7 - (+12) = 7 - 12 = -5$$
$$5 - (-4) = 5 + 4 = 9$$

Rule 5

When multiplying or dividing two numbers with the same sign, the answer is always positive.

$$3 \times 7 = 21 \qquad -8 \bullet -3 = 24 \qquad 12(8) = 96$$

Notice the three different ways of indicating multiplication: an × sign, a "center dot" (•) between the numbers, or no sign at all. Parentheses around an expression just means to treat what's inside as a single number; we'll talk more about that a little later.

$$15 \div 5 = 3 \qquad (-9) \div (-1) = 9 \qquad \frac{-6}{-4} = 1.5$$

Notice the two different ways of indicating division: an ÷ sign or a line in between two numbers.

Rule 6

When multiplying or dividing two numbers with different signs, the answer is always *negative*.

$$3 \times -7 = -21 \qquad -8 \bullet 3 = -24 \qquad (12)(-8) = -96$$

$$-15 \div 5 = -3 \qquad (9) \div (-1) = -9 \qquad \frac{-6}{4} = -1.5$$

Rules 5 and 6 aren't as "sensible" as some of the other rules, and the number line won't help you much with them. Just memorize.

Rule 7

With more than two numbers to be multiplied or divided, take them pairwise, in order, and follow Rules 5 and 6. (The rule for multiplication and division is that if there are an odd number of negative numbers, the answer will be negative; with an even number of negatives, the answer will be positive. If this helps, use it. If not, forget it.)

$$3 \times -2 \times -4 \times 3 = -6 \times -4 \times 3 = 24 \times 3 = 72$$
$$40 \div 2 \div 2 \div -2 = 20 \div 2 \div -2 = 10 \div -2 = -5$$
$$6 \div 3 \times -1 \div 5 = 2 \times -1 \div 5 = -2 \div 5 = -.4$$

FRACTIONS

Rule 1

A fraction is another way of symbolizing division. A fraction means "divide the first (top) expression (the numerator) by the second (bottom) expression (the denominator)." Fractions answer the question, "If I cut (the top number) up into (the bottom number) of equal pieces, how much will be in each piece?"

$$1 \div 2 = .5 \qquad 4 \div 2 = 2 \qquad 4 \div -2 = -2$$

$$\frac{-9}{3} = -3 \qquad \frac{6}{4 - 1} = 2 \qquad \frac{(13 - 3)(7 + 3)}{-3 + 2} = \frac{(10)(10)}{-1} = -100$$

Rule 2

Dividing any number by zero is impossible. If any problem in this book seems to be asking you to divide by zero, you've made an arithmetic mistake somewhere.

Rule 3

You can turn any expression into a fraction by making the original expression the numerator and putting a 1 into the denominator.

$$3 = \frac{3}{1} \qquad -6.2 = \frac{-6.2}{1} \qquad 3x + 4 = \frac{3x + 4}{1}$$

Rule 4

To multiply fractions, multiply their numerators together and multiply their denominators together.

$$\frac{2}{3} \bullet \frac{1}{2} = \frac{2}{6} = .33$$

$$\frac{1}{5} \bullet 10 = \frac{1}{5} \bullet \frac{10}{1} = \frac{10}{5} = 2$$

$$3xy \bullet \frac{3}{4} = \frac{3xy}{1} \bullet \frac{3}{4} = \frac{3(3xy)}{4} = \frac{9xy}{4}$$

Rule 5

Multiplying both the numerator and the denominator of a fraction by the same number doesn't change its value.

$$\frac{1}{2} = \frac{2 \bullet 1}{2 \bullet 2} = \frac{2}{4} = \frac{100}{200} = \frac{\left(\dfrac{1}{200}\right) \bullet 100}{\left(\dfrac{1}{200}\right) \bullet 200} = \frac{.5}{1} = .5$$

Rule 6

To divide by a fraction, invert and multiply. That is, take the fraction you're dividing by (the divisor), switch the denominator and numerator, and then multiply it by the thing into which you're dividing (the dividend).

$$21 \div \frac{3}{5} = \frac{21}{1} \div \frac{3}{5} = \frac{21}{1} \bullet \frac{5}{3} = \frac{105}{3} = 35$$

(I cheated a little in that last example and used some algebra. If you don't understand it yet, come back to it after you've read the "Equations" section of this appendix.)

Rule 7

To add or subtract fractions, they must have a common denominator; that is, their denominators must be the same. For example, you can't add 2/3 and 1/5 as they are. You have to change them to equivalent fractions with a common denominator. How? By multiplying the denominators (and, of course, the numerators) by a number that will make the denominators equal. Of course, you don't have to use the same number for each fraction. Multiply each fraction by the smallest numbers that will make the denominators equal. You may have heard of the "least common denominator": That's what you're looking for. For example, $2/3 = 10/15$ and $1/5 = 3/15$. Then add or subtract the numerators, leaving the denominator unchanged.

$$\frac{2}{3} + \frac{1}{2} = \frac{2 \bullet 2}{2 \bullet 3} + \frac{3 \bullet 1}{3 \bullet 2} = \frac{4}{6} + \frac{3}{6} = \frac{7}{6} = 1\frac{1}{6}$$

$$\frac{1}{5} + \frac{1}{10} = \frac{2 \bullet 1}{2 \bullet 5} + \frac{1 \bullet 1}{1 \bullet 10} = \frac{2}{10} + \frac{1}{10} = \frac{3}{10}$$

$$\frac{5}{8} - \frac{1}{2} = \frac{1 \bullet 5}{1 \bullet 8} - \frac{4 \bullet 1}{4 \bullet 2} = \frac{5}{8} - \frac{4}{8} = \frac{1}{8}$$

$$\frac{1}{3} + \frac{1}{2} + \frac{3}{4} - \frac{1}{12} = \frac{4}{12} + \frac{6}{12} + \frac{9}{12} - \frac{1}{12} = \frac{18}{12} = 1\frac{6}{12} = 1\frac{1}{2}$$

DECIMALS AND PERCENTS

Rule 1

Decimals indicate that the part of the number following the decimal point is a fraction with 10, 100, 1,000, and so on, as the denominator. Rather than trying to find words to express the rule, let me just show you:

$$3.2 = 3\frac{2}{10} \qquad 3.25 = 3\frac{25}{100} \qquad 3.257 = 3\frac{257}{1,000}$$

See how it works? Not particularly complicated, right?

Rule 2

Some fractions, divided out, produce decimals that go on and on and on. To get rid of unneeded decimal places, we can *round off* a number. Say you have the number 1.41421 and you want to express it with just two decimal places. Should your answer be 1.41 or 1.42? The first step in deciding is to create a new number from the digits left over after you take away the ones you want to keep, with a decimal point in front of it. In our example, we keep 1.41, and the newly created number is .421. The next steps are as follows:

1. If the new decimal number is less than .5, just throw it away; you're done with the rounding-off process.
2. If the new decimal is .5 or greater, throw it away, but increase the last digit of the number you keep by 1. For example, 1.41684 would round to 1.42.

Rule 3

Percents are simply fractions of 100 (two-place decimals).

$$45\% = \frac{45}{100} = .45 \qquad 1.3\% = \frac{1.3}{100} = .013 \qquad 110\% = \frac{110}{100} = 1.1$$

EXPONENTS AND ROOTS

Rule 1

An exponent is a small number placed slightly higher than and to the right of a number or expression. For example, 3^3 has an exponent of 3; $(x - y)^2$ has an exponent of 2. An exponent tells how many times a number or expression is to be multiplied by itself.

$$5^2 = 5 \bullet 5 = 25 \qquad 10^3 = 10 \bullet 10 \bullet 10 = 1{,}000 \qquad Y^4 = Y \bullet Y \bullet Y \bullet Y$$

Rule 2

Roots are the opposite of exponents. You can have square roots (the opposite of an exponent of 2), cube roots (opposite of an exponent of 3), and so on. In statistics, we often use square roots, and seldom any other kind, so I'm just going to talk about square roots here.

Rule 3

The square root of a number is the value that, when multiplied by itself, equals that number. For example, the square root of 9 is 3 and $3 \bullet 3 = 9$. The instruction to compute a square root (mathematicians call it "extracting" a square root, but I think that has unfortunate associations to wisdom teeth) is a "radical" sign: $\div$. You take the square root of everything that's shown under the "roof" of the radical.

$$\sqrt{9} = 3 \qquad \sqrt{8100} = 90 \qquad \sqrt{36 + 13} = \sqrt{49} = 7$$
$$\sqrt{36} + 13 = 6 + 13 = 9$$

When extracting a square root, you have three alternatives:

1. Use a calculator with a square root button.
2. Learn how to use a table of squares and square roots.
3. Find a sixth grader who has just studied square roots in school.

I simply cannot recommend alternative (a) too strongly, given the inconvenience of using tables and the unreliability of some sixth graders.

ORDER OF COMPUTATION

Rule 1

When an expression is enclosed in parentheses (like this), treat what's inside like a single number. Do any operations on that expression first, before going on to what's outside the parentheses. With nested parentheses, work from the inside out.

$$4(7 - 2) = 4(5) = 20$$
$$9 \div (-4 + 1) = 9 \div -3 = -3$$
$$12 \times (5 - (6 \times 2)) = 12 \times (5 - 12) = 12 \times -7 = -84$$
$$12 \times (5 - 6) \times 2 = 12 \times (-1) \times 2 = -12 \times 2 = -24$$

With complicated fractions, treat the numerator and the denominator as if each were enclosed in parentheses. In other words, calculate the whole numerator and the whole denominator first, then divide the numerator by the denominator:

$$\frac{3 + 2}{5 \bullet (7 - 3)} = \frac{5}{5 \bullet 4} = \frac{5}{20} = \frac{1}{4}$$

Rule 2

If you don't have parentheses to guide you, do all multiplication and division before you add or subtract.

$$5 + 3 \bullet 2 - 4 = 5 + (3 \bullet 2) - 4 = 5 + 6 - 4 = 7$$
$$8 \div 2 + 9 - 1 - (-2) \bullet (-5) - 5 = 4 + 9 - 1 - (+10) - 5 = -3$$

An algebra teacher in North Dakota taught her students a mnemonic to help them remember the correct order of operation: My Dear Aunt Sally → Multiply, Divide, Add, Subtract.

Rule 3

Exponents and square roots are treated as if they were a single number. That means you square numbers or take square roots first of all—before adding, subtracting, multiplying, or dividing. Maybe we should treat My Dear Aunt Sally like a mean landlord and make the rule be "Roughly Evict My Dear Aunt Sally," in order to get the roots and exponents first in line!

Here are some examples of how the order of computation rules work together:

$$5 - (3 \times 4) \times (8 - 2^2)(-3 + 1) \div 3$$
$$= 5 - (3 \times 4) \times (8 - 4)(-3 + 1) \div 3 \quad \text{(exponent)}$$
$$= 5 - (12) \times (4) \times (-2) \div 3 \quad \text{(things inside parentheses)}$$
$$= 5 - 48 \times -2 \div 3 \quad \text{(multiply)}$$
$$= 5 - (-96) \div 3 \quad \text{(multiply again)}$$
$$= 5 - (-32) \quad \text{(divide)}$$
$$= 37 \quad \text{(and the addition comes last)}$$

Did you remember that subtracting a negative number is the same as adding a positive number?

$$2x - 3^2 \div (3 + 2) - \sqrt{25} \bullet 10 + (8 - (3 + 4))$$
$$= 2x - 9 \div (3 + 2) - 5 \bullet 10 + (8 - (3 + 4))$$
$$= 2x - 9 \div 5 - 5 \bullet 10 + (8 - 7)$$
$$= 2x - 9 \div 5 - 5 \bullet 10 + 1$$
$$= 2x - 1.8 - 50 + 1$$
$$= 2x - 50.8$$

SUMMATION

A summation sign looks like a goat's footprint: Σ. Its meaning is pretty simple—add up what comes next. Most of the time, "what comes next" is obvious from the context. If you have a variable designated as x, with individual values x_1, x_2, x_3, and so on, then Σx refers to the sum of all those individual values.

Actually, Σx is a shorthand version of $\Sigma_{i=1}^{N} x$, which means that there are N individual x's. Each x is called x_i, and the values of i run from 1 to N. When $i = 10$ and $N = 50$, x_i would be the tenth in a set of 50 variables; $\Sigma_{i=1}^{N} x$ would mean to find the sum of all 50 of them. For our purposes, a simple $\Sigma_{i=1} x$ says the same thing and we'll just use that.

There are a few rules that you should know about doing summation, however. Let's look at an example. Five people take a pop quiz, and their scores are 10, 10, 8, 12, and 10. In other words, $x_1 = 10$, $x_2 = 10$, $x_3 = 8$, $x_4 = 12$, and $x_5 = 10$. $\Sigma x = 50$. What about Σx^2? Well, that would be $100 + 100 + 64 + 144 + 100$. $\Sigma x^2 = 508$.

Now, does it make sense to you that $\Sigma x^2 \neq (\Sigma x)^2$? This is a key idea, and it has to do with the order of computation. $(\Sigma x)^2$ is read "sum of x, quantity squared," and the parentheses mean that you add up all the x's first and square the sum: $(\Sigma x)^2 = (50)^2 = 2500$.

Now, just to make things interesting, we'll throw in another variable. Let y stand for scores on another quiz: $y_1 = 4$, $y_2 = 5$, $y_3 = 6$, $y_4 = 5$, $y_5 = 4$. $\Sigma y = 24$, $\Sigma y^2 = 118$, and $(\Sigma y)^2 = 576$. And we have some new possibilities:

$$\Sigma x + \Sigma y \qquad \Sigma x^2 + \Sigma y^2 \qquad \Sigma(x + y)$$
$$\Sigma(x^2 + y^2) \qquad \Sigma(x + y)^2 \qquad (\Sigma(x + y))^2$$

See if you can figure out these values on your own, and then we'll go through each one.

$\Sigma x + \Sigma y$ — Add up the x values, add up the y values, add them together: 74.

$\Sigma x^2 + \Sigma y^2$ — Add up the squared x values, add up the squared y values, add them together: 626.

$\Sigma(x + y)$ — Add each x, y pair together, and add up the sums:

$14 + 15 + 14 + 17 + 14 = 74$. Yes, $\Sigma x + \Sigma y = \Sigma(x + y)$. Every time.

$\Sigma(x^2 + y^2)$ — Square an x and add it to its squared y partner; then add up the sums:

$116 + 125 + 100 + 169 + 116 = 626$. $\Sigma(x^2 + y^2) = \Sigma x^2 + \Sigma y^2$.

$\Sigma(x + y)^2$ — Add each x, y pair together, square the sums, and add them up: 1,102.

$(\Sigma(x + y))^2$ — Did those double parentheses throw you? Use them like a road map, to tell you where to go first. Starting from the inside, you add each x, y pair together. Find the sum of the pairs, and last of all square that sum: 5,476.

EQUATIONS

An equation is two expressions joined by an equal sign. Not surprisingly, the value of the part in front of the equal sign is exactly equal to the value of the part after the equal sign.

Rule 1

Adding or subtracting the same number from each side of an equation is acceptable; the two sides will still be equivalent.

$5 + 3 = 9 - 1$	$5 + 3 + 5 = 9 - 1 + 5$	$5 + 3 - 5 = 9 - 1 - 5$
$8 = 8$	$13 = 13$	$3 = 3$
$6 \div 4 + 1 \div 2 = 2$	$6 \div 4 + 1 \div 2 + 5 = 2 + 5$	$6 \div 4 + 1 \div 2 - 5 = 2 - 5$
$2 = 2$	$7 = 7$	$-3 = -3$
$12 - 2 = (2)(5)$	$12 - 2 + 5 = (2)(5) + 5$	$12 - 2 - 5 = (2)(5) - 5$
$10 = 10$	$15 = 15$	$5 = 5$

Rule 2

If you add or subtract a number from one side of an equation, you must add or subtract it from the other side as well if the equation is to balance, that is, if both sides are to remain equal.

$$8 - 2 = 3 + 3 \quad 8 - 2 + 2 = 3 + 3 + 2 \quad 8 = 8$$
$$2x + 7 = 35 \quad 2x + 7 - 7 = 35 - 7 \quad 2x = 28$$

Rule 3

If you multiply or divide one side of an equation by some number, you must multiply or divide the other side by the same number. You can't multiply or divide just one part of each side; you have to multiply or divide the whole thing.

$$3 + 2 - 1 = 7 - 5 + 2$$

Multiply both sides by 6:

$$6 \bullet (3 + 2 - 1) = 6 \bullet (7 - 5 + 2)$$
$$6 \bullet (4) = 6 \bullet (4)$$
$$24 = 24$$

Look what would happen if you multiplied just one of the numbers on each side by 6:

$$6 \bullet (3) + 2 - 1 = 6 \bullet (7) - 5 + 2$$
$$18 + 2 - 1 = 42 - 5 + 2$$
$$19 = 39$$

Writing out these kinds of rules is a lot like eating hot buttered popcorn: It's hard to know when to quit. And, as with popcorn, it's a lot better to quit too soon than to quit too late; the former leaves you ready for more tomorrow, while the latter can make you swear off the stuff for months.

We could go on and on here, and end up with the outline for a freshman math course, but that's not our purpose. These rules will allow you to do all the math in this book and a great deal of the math in more advanced statistics courses. So let's get going on the fun part!

Appendix B

Proportions of Area Under the Standard Normal Curve

z	0 z	0 z	z	0 z	0 z	z	0 z	0 z
0.00	.0000	.5000	0.17	.0675	.4325	0.34	.1331	.3669
0.01	.0040	.4960	0.18	.0714	.4286	0.35	.1368	.3632
0.02	.0080	.4920	0.19	.0753	.4247	0.36	.1406	.3594
0.03	.0120	.4880	0.20	.0793	.4207	0.37	.1443	.3557
0.04	.0160	.4840	0.21	.0832	.4168	0.38	.1480	.3520
0.05	.0199	.4801	0.22	.0871	.4129	0.39	.1517	.3483
0.06	.0239	.4761	0.23	.0910	.4090	0.40	.1554	.3446
0.07	.0279	.4721	0.24	.0948	.4052	0.41	.1591	.3409
0.08	.0319	.4681	0.25	.0987	.4013	0.42	.1628	.3372
0.09	.0359	.4641	0.26	.1026	.3974	0.43	.1664	.3336
0.10	.0398	.4602	0.27	.1064	.3936	0.44	.1700	.3300
0.11	.0438	.4562	0.28	.1103	.3897	0.45	.1736	.3264
0.12	.0478	.4522	0.29	.1141	.3859	0.46	.1772	.3228
0.13	.0517	.4483	0.30	.1179	.3821	0.47	.1808	.3192
0.14	.0557	.4443	0.31	.1217	.3783	0.48	.1844	.3156
0.15	.0596	.4404	0.32	.1255	.3745	0.49	.1879	.3121
0.16	.0636	.4364	0.33	.1293	.3707	0.50	.1915	.3085

(continued)

Source: Runyon and Haber, *Fundamentals of Behavioral Statistics*, 2nd ed., 1971, Addison-Wesley, Reading, Mass.

APPENDIX B (continued)

z	0 z	0 z	z	0 z	0 z	z	0 z	0 z
0.51	.1950	.3050	0.89	.3133	.1867	1.27	.3980	.1020
0.52	.1985	.3015	0.90	.3159	.1841	1.28	.3997	.1003
0.53	.2019	.2981	0.91	.3186	.1814	1.29	.4015	.0985
0.54	.2054	.2946	0.92	.3212	.1788	1.30	.4032	.0968
0.55	.2088	.2912	0.93	.3238	.1762	1.31	.4049	.0951
0.56	.2123	.2877	0.94	.3264	.1736	1.32	.4066	.0934
0.57	.2157	.2843	0.95	.3289	.1711	1.33	.4082	.0918
0.58	.2190	.2810	0.96	.3315	.1685	1.34	.4099	.0901
0.59	.2224	.2776	0.97	.3340	.1660	1.35	.4115	.0885
0.60	.2257	.2743	0.98	.3365	.1635	1.36	.4131	.0869
0.61	.2291	.2709	0.99	.3389	.1611	1.37	.4147	.0853
0.62	.2324	.2676	1.00	.3413	.1587	1.38	.4162	.0838
0.63	.2357	.2643	1.01	.3438	.1562	1.39	.4177	.0823
0.64	.2389	.2611	1.02	.3461	.1539	1.40	.4192	.0808
0.65	.2422	.2578	1.03	.3485	.1515	1.41	.4207	.0793
0.66	.2454	.2546	1.04	.3508	.1492	1.42	.4222	.0778
0.67	.2486	.2514	1.05	.3531	.1469	1.43	.4236	.0764
0.68	.2517	.2483	1.06	.3554	.1446	1.44	.4251	.0749
0.69	.2549	.2451	1.07	.3577	.1423	1.45	.4265	.0735
0.70	.2580	.2420	1.08	.3599	.1401	1.46	.4279	.0721
0.71	.2611	.2389	1.09	.3621	.1379	1.47	.4292	.0708
0.72	.2642	.2358	1.10	.3643	.1357	1.48	.4306	.0694
0.73	.2673	.2327	1.11	.3665	.1335	1.49	.4319	.0681
0.74	.2704	.2296	1.12	.3686	.1314	1.50	.4332	.0668
0.75	.2734	.2266	1.13	.3708	.1292	1.51	.4345	.0655
0.76	.2764	.2236	1.14	.3729	.1271	1.52	.4357	.0643
0.77	.2794	.2206	1.15	.3749	.1251	1.53	.4370	.0630
0.78	.2823	.2177	1.16	.3770	.1230	1.54	.4382	.0618
0.79	.2852	.2148	1.17	.3790	.1210	1.55	.4394	.0606
0.80	.2881	.2119	1.18	.3810	.1190	1.56	.4406	.0594
0.81	.2910	.2090	1.19	.3830	.1170	1.57	.4418	.0582
0.82	.2939	.2061	1.20	.3849	.1151	1.58	.4429	.0571
0.83	.2967	.2033	1.21	.3869	.1131	1.59	.4441	.0559
0.84	.2995	.2005	1.22	.3888	.1112	1.60	.4452	.0548
0.85	.3023	.1977	1.23	.3907	.1093	1.61	.4463	.0537
0.86	.3051	.1949	1.24	.3925	.1075	1.62	.4474	.0526
0.87	.3078	.1922	1.25	.3944	.1056	1.63	.4484	.0516
0.88	.3106	.1894	1.26	.3962	.1038	1.64	.4495	.0505

APPENDIX B (continued)

z	0 z	0 z	z	0 z	0 z	z	0 z	0 z
1.65	.4505	.0495	2.03	.4788	.0212	2.41	.4920	.0080
1.66	.4515	.0485	2.04	.4793	.0207	2.42	.4922	.0078
1.67	.4525	.0475	2.05	.4798	.0202	2.43	.4925	.0075
1.68	.4535	.0465	2.06	.4803	.0197	2.44	.4927	.0073
1.69	.4545	.0455	2.07	.4808	.0192	2.45	.4929	.0071
1.70	.4554	.0446	2.08	.4812	.0188	2.46	.4931	.0069
1.71	.4564	.0436	2.09	.4817	.0183	2.47	.4932	.0068
1.72	.4573	.0427	2.10	.4821	.0179	2.48	.4934	.0066
1.73	.4582	.0418	2.11	.4826	.0174	2.49	.4936	.0064
1.74	.4591	.0409	2.12	.4830	.0170	2.50	.4938	.0062
1.75	.4599	.0401	2.13	.4834	.0166	2.51	.4940	.0060
1.76	.4608	.0392	2.14	.4838	.0162	2.52	.4941	.0059
1.77	.4616	.0384	2.15	.4842	.0158	2.53	.4943	.0057
1.78	.4625	.0375	2.16	.4846	.0154	2.54	.4945	.0055
1.79	.4633	.0367	2.17	.4850	.0150	2.55	.4946	.0054
1.80	.4641	.0359	2.18	.4854	.0146	2.56	.4948	.0052
1.81	.4649	.0351	2.19	.4857	.0143	2.57	.4949	.0051
1.82	.4656	.0344	2.20	.4861	.0139	2.58	.4951	.0049
1.83	.4664	.0336	2.21	.4864	.0136	2.59	.4952	.0048
1.84	.4671	.0329	2.22	.4868	.0132	2.60	.4953	.0047
1.85	.4678	.0322	2.23	.4871	.0129	2.61	.4955	.0045
1.86	.4686	.0314	2.24	.4875	.0125	2.62	.4956	.0044
1.87	.4693	.0307	2.25	.4878	.0122	2.63	.4957	.0043
1.88	.4699	.0301	2.26	.4881	.0119	2.64	.4959	.0041
1.89	.4706	.0294	2.27	.4884	.0116	2.65	.4960	.0040
1.90	.4713	.0287	2.28	.4887	.0113	2.66	.4961	.0039
1.91	.4719	.0281	2.29	.4890	.0110	2.67	.4962	.0038
1.92	.4726	.0274	2.30	.4893	.0107	2.68	.4963	.0037
1.93	.4732	.0268	2.31	.4896	.0104	2.69	.4964	.0036
1.94	.4738	.0262	2.32	.4898	.0102	2.70	.4965	.0035
1.95	.4744	.0256	2.33	.4901	.0099	2.71	.4966	.0034
1.96	.4750	.0250	2.34	.4904	.0096	2.72	.4967	.0033
1.97	.4756	.0244	2.35	.4906	.0094	2.73	.4968	.0032
1.98	.4761	.0239	2.36	.4909	.0091	2.74	.4969	.0031
1.99	.4767	.0233	2.37	.4911	.0089	2.75	.4970	.0030
2.00	.4772	.0228	2.38	.4913	.0087	2.76	.4971	.0029
2.01	.4778	.0222	2.39	.4916	.0084	2.77	.4972	.0028
2.02	.4783	.0217	2.40	.4918	.0082	2.78	.4973	.0027

(*continued*)

APPENDIX B (continued)

z	0 z	0 z	z	0 z	0 z	z	0 z	0 z
2.79	.4974	.0026	2.98	.4986	.0014	3.17	.4992	.0008
2.80	.4974	.0026	2.99	.4986	.0014	3.18	.4993	.0007
2.81	.4975	.0025	3.00	.4987	.0013	3.19	.4993	.0007
2.82	.4976	.0024	3.01	.4987	.0013	3.20	.4993	.0007
2.83	.4977	.0023	3.02	.4987	.0013	3.21	.4993	.0007
2.84	.4977	.0023	3.03	.4988	.0012	3.22	.4994	.0006
2.85	.4978	.0022	3.04	.4988	.0012	3.23	.4994	.0006
2.86	.4979	.0021	3.05	.4989	.0011	3.24	.4994	.0006
2.87	.4979	.0021	3.06	.4989	.0011	3.25	.4994	.0006
2.88	.4980	.0020	3.07	.4989	.0011	3.30	.4995	.0005
2.89	.4981	.0019	3.08	.4990	.0010	3.35	.4996	.0004
2.90	.4981	.0019	3.09	.4990	.0010	3.40	.4997	.0003
2.91	.4982	.0018	3.10	.4990	.0010	3.45	.4997	.0003
2.92	.4982	.0018	3.11	.4991	.0009	3.50	.4998	.0002
2.93	.4983	.0017	3.12	.4991	.0009	3.60	.4998	.0002
2.94	.4984	.0016	3.13	.4991	.0009	3.70	.4999	.0001
2.95	.4984	.0016	3.14	.4992	.0008	3.80	.4999	.0001
2.96	.4985	.0015	3.15	.4992	.0008	3.90	.49995	.00005
2.97	.4985	.0015	3.16	.4992	.0008	4.00	.49997	.00003

Appendix C

....

Random Selection

USING THE RANDOM-NUMBER TABLE TO DRAW A SAMPLE

Step 1

Define your population, for example, all the first-grade students in the six elementary schools in Gainesville, Florida.

Step 2

List all the members of the population. In our example, you would have to go to the individual schools or to the Board of Education and get this information. (This is the hardest step.)

Step 3

Assign an identification number to each member of the population: 1, 2, 3, 4, 5, and on out through the last student on your list. Enter these numbers into the Data Editor of SPSS as shown in Figure C-1.

Step 4

Decide on the size of your sample. This will depend on all sorts of things: the kind of experiment you plan to do, the consequences of drawing a wrong conclusion (the likelihood of error goes down as the sample size goes up), the amount of money available. Let's say you decide to draw a sample of 5 from the population of 10.

Step 5

Click on *Data* in the menu bar and then on *Select Cases . . .* to open the dialog box shown in Figure C-2.

In the *Select Cases* dialog box, select *Random sample of cases* and then click on *Sample* to open the *Select Cases: Random Sample* dialog box shown in Figure C-3.

You can select a random sample in two ways in SPSS. You can select a specified percentage of cases or the exact number desired. For our example, select *Exactly* and then type in 5 and 10, as shown in the figure, and click *Continue* and then *OK*. Figure C-4 shows the results of our random selection.

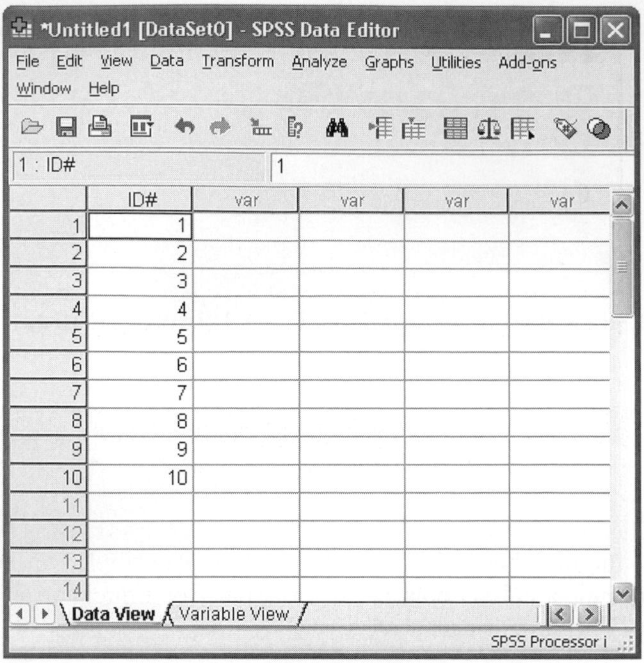

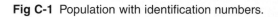

Fig C-1 Population with identification numbers.

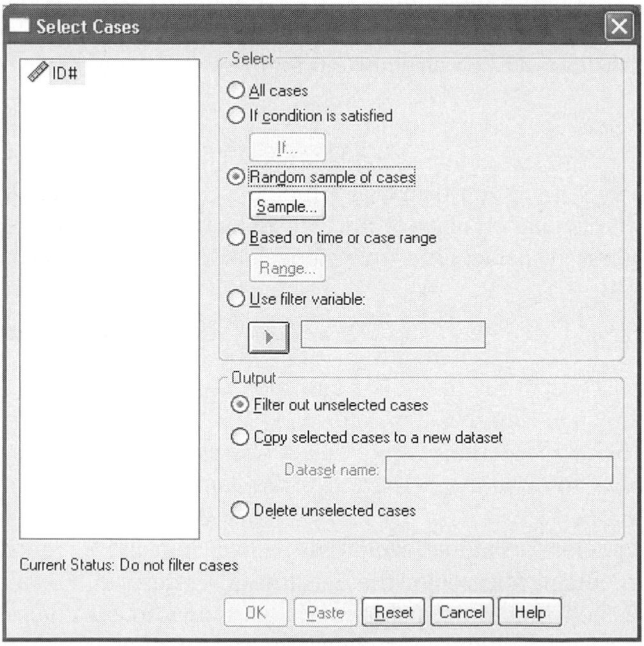

Fig C-2 Select cases dialog box.

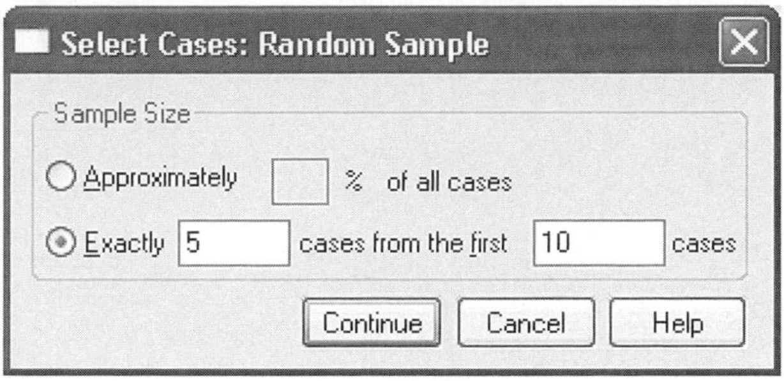

Select Cases: Random Sample ☒

Sample Size

○ Approximately [   ] % of all cases

◉ Exactly [5] cases from the first [10] cases

[Continue] [Cancel] [Help]

Fig C-3 Select cases: Random sample dialog box.

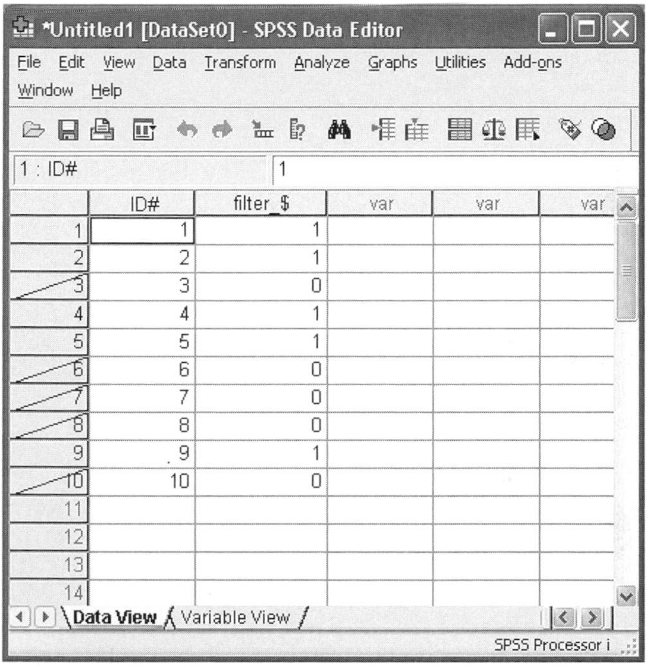

⌂ *Untitled1 [DataSet0] - SPSS Data Editor ▭ ▢ ☒

File Edit View Data Transform Analyze Graphs Utilities Add-ons
Window Help

1 : ID#   1

	ID#	filter_$	var	var	var
1	1	1			
2	2	1			
3	3	0			
4	4	1			
5	5	1			
6	6	0			
7	7	0			
8	8	0			
9	.9	1			
10	10	0			
11					
12					
13					
14					

◀ ▶ \ **Data View** ⟨ Variable View /   ◁ ▷

SPSS Processor i

Fig C-4 Random selection.

As you can see in the figure, SPSS randomly selected 5 of the 10 cases for participation in the study and crossed out the cases that were not selected. If you select the *Deleted* option in the *Unselected Cases Are* area, then the unselected cases are deleted and you're presented with the list of identification numbers for the randomly sampled participants. Alternatively, you can select to have the selected cases copied to a new dataset, which would leave your original dataset intact. Either way, it's very easy to get a random sample using SPSS.

INDEX